BUSINESS FUNDAMENTALS FOR EN

BUSINESS FUNDAMENTALS FOR ENGINEERS

Chengi Kuo

University of Strathclyde

McGRAW-HILL BOOK COMPANY

London · New York · St Louis · San Francisco · Auckland
Bogotá · Caracas · Hamburg · Lisbon · Madrid · Mexico
Milan · Montreal · New Delhi · Panama · Paris · San Juan
São Paulo · Singapore · Sydney · Tokyo · Toronto

Published by
McGRAW-HILL Book Company Europe
Shoppenhangers Road, Maidenhead, Berkshire, SL6 2QL, England
Tel 0628 23432; Fax 0628 770224

British Library Cataloguing in Publication Data

Kuo, Chengi
  Business fundamentals for engineers.
  I. Title
  620.0068

  ISBN 0-07-707423-8

Library of Congress Cataloging-in-Publication Data

Kuo, Chengi.
  Business fundamentals for engineers / Chengi Kuo.
    p.   cm.
  Includes bibliographical references and index.
  ISBN 0-07-707423-8
  1. Engineering–Management. I. Title
  TA190.K86 1992   91-28253
  620'0068–dc20   CIP

12345 CL 95432

Typeset by Wyvern Typesetting
and printed and bound in Great Britain by Clays Ltd, St Ives plc

This work is dedicated to
Toh Yong for her understanding
support and affection

CONTENTS

FOREWORD

Sir Graham Day

Business Fundamentals For Engineers makes two important primary contributions. The book introduces young engineers to business, the environment within which most will practice their profession. Also, it will assist the conversion of engineers into business managers, an inevitable development for many. But that is not all that *Business Fundamentals* does.

Engineering is no different from other professions, such as law and accounting, that operate both on the margins of business and within business. To individual and professional detriment, all too frequently many such professionals fail to see their contributions in the fuller business context. Thus, while Professor Chengi Kuo's book is intended to provide engineers with an understanding of business, I am convinced that lawyers (I am one) and accountants can benefit as well.

However, even the professional groups may be too narrow a target audience. Professor Kuo deals with markets and how they operate, economics and accounting, personnel and communications issues and, even, government. From my perspective, *Business Fundamentals* justifies being included in curricula for a wide variety of students.

For post-secondary education, many students select engineering as the logical extension of a preference for and a grounding in quantitative subjects. Many business subjects being fundamentally qualitative, are thus unattractive. *Business Fundamentals For Engineers* uniquely uses a quantitative type presentational technique that enhances the text's appeal and utility, particularly for people with a quantitative bias.

While I believe the book's initial and primary appeal will be to engineering students and recent graduate engineers, I consider that many well-established engineers could benefit materially by reading it. Perhaps, for the latter, it should be available also in a plain brown wrapper!

For me *Business Fundamentals For Engineers* has a particular appeal. It is useful. Too many academics publish because they must. Others, understandably, write for a peer group audience. A precious few research and write so as to contribute to the third party process of learning. Professor Kuo, a polymath if ever there was one, is a committed teacher. This book amply supports his vocation. The teacher shows through in the use of the case studies and a helpful glossary. To research and write is good, to teach is divine.

Chengi Kuo's readers will be fortunate in having such a book written by a man with a very wide range of interests and experience. For my part, I am honoured to be his friend and am very pleased to have had the opportunity to write this brief foreword.

PREFACE

For a number of years I have found that the efficiency and effectiveness of my academic and technological activities and my wide dealings with the industrial and commercial world are enhanced by having acquired an understanding of, and achieving some working skill in, various aspects of what is commonly called 'business' or 'management'. Some years ago, because of my own experience, I began to give our students active encouragement to take classes in one or two business subjects such as Marketing, Economics and Communications Skills. I was, therefore, somewhat disappointed when only a few of them shared my enthusiasm for these subjects. Discussing the matter with different groups of students, however, enabled me to identify three problems.

Firstly, being students on engineering courses they tended to concentrate mainly on technological topics and had difficulty in adjusting to the less numeric treatment and less precise methods of many non-technical subjects. Secondly, these students were unfamiliar with the methods of studying such subjects and were not very efficient at preparing answers to questions based on extensive reading of recommended texts. Thirdly, and perhaps most importantly, the majority of them failed to grasp the basic principles of the subjects and had a very narrow view of their scope and relevance. Consequently, they found it difficult to relate these subjects to their principal courses.

The most positive response came from one small group of students who commented that they could see the relevance of business subjects but had difficulty in applying the knowledge in practice. In other words, they were failing to acquire the basic understanding of these subjects which would enable them to gain access to useful business knowledge as a tool. It was in response to this suggestion that I took up the challenge to build a 'foundation' for them by providing a 'bridging' text.

When I began to examine the books and other publications already available on business and to discuss these with colleagues and contacts, two factors soon became obvious. The first was that the subject scope is very broad, and the second that the published material has tended to specialize in specific topics, such as Marketing or Finance. It is only in recent years that attempts have been made to provide books which cover a broad range of topics within a single volume and, although interesting, most of them were around six hundred pages long. In some cases the text had been written by a number of authors each of whom treated their section as a special subject only loosely connected with business as a whole. I was unable to find a single book which was concise, introductory and yet fulfilled the interpretive function that I had perceived as necessary. There was, therefore, a potential market for a short- to medium-length introductory text which would enable someone with a technological training to gain a sound appreciation of business fundamentals.

This book has been written as if it were a technological text, with the principles introduced in 24 sections and their application demonstrated by 72 illustrative examples. Three major case examples were also included to show how the various aspects of business are integrated in practical situations.

It has been quite a challenge to find a range of suitable examples. Those used have been drawn from a mixture of referenced publications, private communications, discussions at conferences and seminars, and press comments. The last-named source comprises both technical publications and newspaper reports and articles.

It is my hope that a study of this book will enable readers to enhance their technological activities by developing efficient methods of applying business thinking and techniques in this area.

The book could not have been written without the help, support and encouragement of many friends, colleagues and students. I should like to acknowledge my grateful thanks, in particular to the following: Bill Bryce, Dorothy Cameron, David Crowther, Nick Currie, Ingolf Grinde, John Harwood, Jeremy Haslam, Gordon Hayward, Barry Koch, Yong Luo, Jim McCabe, Jim McGilvary, Lewis McIver, Harry Osborne, Jim Paterson, Ben Pedret, Elizabeth Proudfoot, Yahaya Sanusi, Bill Scott, Bob Scott and Peter Williams. Special thanks are due to Christine Hutcheon for assisting me in the preparation of the text.

Chengi Kuo

CHAPTER 1

BUSINESS AND
THE ENGINEER

1.1 WHAT IS BUSINESS?

'Business' is the term normally used to describe those activities which together produce goods or services to meet customers' demands in order to make a profit in doing so. These activities are usually carried out through a formally constituted organization with legal status and a recognized identity, such as a company, partnership or institution. Directly or indirectly they will include finding out what customers' needs are, making the most effective use of both human and financial resources, selecting appropriate technologies, planning, processing information, problem-solving and motivating the personnel of the organization.

The basic concepts of business functions can usually be readily understood, but when they interact together the thinking behind them becomes less obvious. The problem is compounded by the fact that many books on business tend to concentrate on detail while paying little or no attention to the basic principles and the key issues. This makes it even more difficult for those not specializing in the subject to acquire a proper understanding of business.

In practice, what someone understands to be 'business' will depend to a large extent on his or her point of view and experience. This is best illustrated by the way in which the following two manufacturing organizations are run. Although both companies are of similar size and provide a similar range and value of products, the backgrounds of their leaders are very different.

1

One leader was trained in engineering and obtained a higher degree qualification in manufacturing technology. For him, business is all about design, production planning, efficiency, quality assurance and computerized applications of machine tools. This, in turn, is reflected in the composition of his team, which is dominated by those with technological training. The Sales Division is represented, but has little influence on company policy. The argument usually put forward to justify his philosophy can be stated as follows:

> Unless the products we manufacture are top quality no one will want to buy them.

The other leader's qualifications are in finance and marketing. For him business is all about clients, prices, market share, profits, generating income, cost control and balancing accounts. This is also reflected in the way decisions are made and in the company's particular emphases. The technical personnel have little influence on the direction of the company and are expected to concentrate on providing products to specification and on schedule. The justification for this approach can be stated as follows:

> The company can only thrive in a changing world if products are sold and income is generated.

These two examples highlight the two extremes in interpreting the term 'business', and clearly a more balanced approach is required if an organization is to succeed in making and maintaining profits. A wider definition of businesss would be:

> The activities which together provide goods, services or some other output for a specific objective, such as to make a profit, fulfil a need or gain some benefit.

On the basis of this definition, business can cover the total spectrum of activities, ranging from engineering and manufacturing through distribution and services to training and an individual's daily work. Thus it is essential for everyone to have an appreciation of business and its associated fundamentals; see Fig. 1.1.

1.2 THE NEED TO STUDY BUSINESS

Having accepted a definition for the term 'business', it would be helpful to ask why an engineer should study business subjects. The question is a valid one, because many of those involved in engineering chose this field out of a desire to be associated with technology, and see business topics—although very interesting—simply as a 'distraction'. Alternatively, they regard business topics as something to be postponed until they reach managerial status! However, when the question is put to people already involved in engineering activities their replies can usually be summarized in one of four comments:

- You need to know something about business if you want to get on.
- We can learn some business skills and techniques.
- They can help to improve performance in our competitive world.
- The knowledge gained can provide an overview of the activities of an organization.

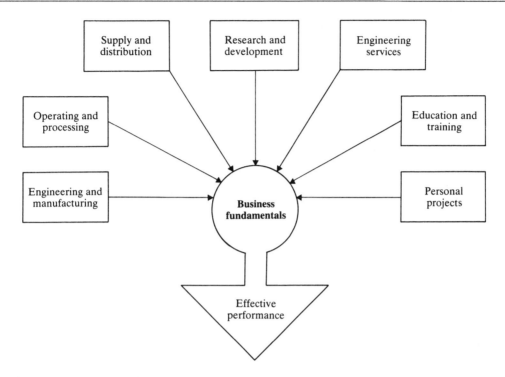

Figure 1.1   Main activities served by business fundamentals

All of these are valid reasons for studying business topics, but there are several other and more compelling reasons why these topics are important for an engineer and why a good understanding of the fundamental elements of business is crucial. The following are the key factors.

To become a more effective engineer

Engineers study many subjects in the course of their education and training. For example, all engineering courses include classes in mathematics and science. The purpose of these is not to produce mathematicians, scientists or engineering mathematicians, but to teach skills that are thought to be useful in support of engineering activities. The same argument also applies to the study of business subjects. An appreciation of the fundamental elements of business and the way they are applied will make engineers more effective, both in their work and other wider activities.

To appreciate the interdependence of subject areas

Technology does not stand on its own, but has to function closely with many other areas of knowledge, including business. This is illustrated by the fact that an engineer can design and build a product, but needs assistance with other tasks. These would include, for example, identifying and defining the potential user's requirements, obtaining finance for research and development work, managing resources and selling the product.

To identify needs more accurately

One of the main roles of the engineer is to transform expressed requirements into reality. This is a challenging task which often involves overcoming a number of technological hurdles and commercial constraints. An understanding of business, and particularly of marketing and market research, will better enable the engineer to employ commercial criteria for selecting the most suitable technical options. In this way needs can be more accurately interpreted through personal contact, and technological effort can be more fruitfully directed.

To improve the engineer's capabilities

The approaches used in treating business topics often differ from those employed in teaching scientific and technological subjects. The engineer will benefit greatly from exposure to these different approaches, particularly when required solutions are not numerically precise or when more than one possibility would be satisfactory. Potential gains include a widening of horizons and improved communication between engineers themselves and with others with whom they come into contact. This is an additional way of enhancing the image of engineers in countries in which they have yet to reach a prestigious position!

To make better use of resources

An engineer's work nearly always involves the deployment of both human and financial resources. This requirement is often embodied in a specification that includes constraints. A knowledge of business will enable the engineer to establish the best interface between technological and commercial aspects. This in turn will contribute towards achieving optimum results.

1.3 THE BASIC KNOWLEDGE

Business involves many activities and deals with a large number of topics. These range from marketing and finance on the one hand to management techniques and personal attributes on the other. Without devoting an unrealistic amount of time to these subjects it would be impossible for an engineer to acquire a working expertise in the complete range. So, how much business knowledge should an engineer acquire? Views differ widely, because one of the prime aims of an engineer's education is to develop a high level of technological competence. Thus too great a focus on business topics could adversely affect the successful meeting of this objective.

On the other hand, too general an introduction will lack impact and will fail to provide an adequate understanding of business. A good case can be made for providing a comprehensive study of one or two key elements of business, such as marketing or project management. Whatever level of business knowledge is considered desirable, two factors must be borne in mind. Firstly, there is a limit to the amount of time that can be devoted to non-technological subjects in an engineering education programme. Secondly, conventional teaching of technological subjects differs greatly from the approach employed for non-technological subjects. In this area, therefore, engineering students need to develop fresh skills rather than simply acquire new knowledge.

For these reasons it is desirable for engineers to gain a broad appreciation of business in its entirety. This should be coupled with a good understanding of the fundamentals, together with their practical application. With this basis the individual will be able to enhance knowledge of, and acquire skill in, any particular aspect of business as and when required.

1.4 FINDING A SUITABLE APPROACH

Having accepted the genuine need for an engineer to have a sound understanding of business, we still face the crucial question: 'How should the fundamentals of business be acquired?'.

There is no shortage of textbooks on business subjects but they tend to specialize in particular areas and are often written in a form unfamiliar to those who have concentrated on engineering and science subjects. The text in many cases is too detailed for an engineer's needs, and the terminology is often abstruse.

There is, therefore, a need for a book that will present business topics in a format similar to that often used for technological material. The text should be written concisely, with each topic selected occupying only a few pages. Sketches and other illustrations should be used in explaining or emphasizing their background and key issues. To assist reading, a glossary should be provided with definitions of key terms.

This book aims to fulfil the above requirements. It begins with a brief analysis of basic business activities and presentation of the four key elements of business. These elements, divided into 'components', are then dealt with in individual chapters. The role of each component is defined, the subject outlined, key issues are considered and illustrative examples are provided to relate the subject to an engineering context. Material for further study is listed at the end of each chapter.

No attempt is made to cover every aspect of the subject, but it is hoped that, having gained an understanding of the fundamental principles and their application, readers will have sufficient interest and confidence to extent their knowledge of specific topics by reading some of the books and journals recommended.

CHAPTER 2

FUNDAMENTAL ELEMENTS OF BUSINESS

2.1 KEY ACTIVITIES

As already indicated, providing goods, services or any other output involves integrating a number of different activities. As an example, it would be useful to identify the activities which are necessary if an organization is to be successful in making and selling a product, be it earth-moving equipment, a sewing machine or an electronic device. Careful study has yielded the following list, with items from both the technological and the business aspects:

- Needs — Identify what potential customers require
- Clients — Sell the design concept or product to customers
- Finance — Raise the necessary finance for the project
- Facilities — Acquire or rent workspace and install the necessary facilities in terms of machinery, office equipment, computer hardware and software
- Agreement — Reach agreement on specifications, delivery dates, cost, methods of payment etc.
- Design — Provide detailed information so that the product can be manufactured
- Planning — Prepare a detailed workplan to ensure that the product will be ready within the agreed time-scale
- Purchase — Obtain equipment and raw materials from suppliers
- Manufacturing — Make the product from the design information via the use of the facilities

- Problem-solving   Seek to overcome organizational problems which may arise
- Personnel   Bring together all the skills and expertise required for the tasks involved
- Control   Maintain manufacturing progress, and control the financial outlay at each stage of the manufacturing process
- Commissioning   Test the finished product to ensure it meets the agreed specifications and quality
- Delivery   Deliver a product that satisfies the client's requirements
- Maintenance   Provide a planned programme for the maintenance of the product over a specified period
- Research   Carry out research and development work to solve problems that have shown up in earlier products and to keep ahead of competition

The interdependence of technology and business can be even more closely identified if the above list of activities is divided into the following three groups:

- *Technological aspects*: Design, facilities, planning, manufacturing, control (production), commissioning, delivery, maintenance and research (the actual work).
- *Business aspects*: Needs, clients, finance, control (costs), problem-solving, agreement, purchasing, personnel.
- *Technological and business aspects*: Needs, facilities, planning, maintenance, research (policy).

Even after this regrouping, the manufacture of any product calls for a combination of business and technological activities. In practice, approaches vary between organizations and according to whether they are producing 'one-offs' or a run of the same item. Approaches also vary between organizations already in business and companies created for a specific purpose. It is also possible to subcontract certain activities to other organizations.

2.2 THE COMMON FACTORS

Although running a business involves many different types of activity, closer examination will yield certain factors common to most, if not all, of them. There are four in particular. Firstly, 'people' are involved in all these activities, from specialists in both technology and management to members of the workforce operating machines and assembling components. The ability to work together plays a crucial role in an organization's effectiveness.

Secondly, financial resources are needed for most of them, to pay the salaries of the staff, cover the cost of facilities, purchase materials, fund research, and so on.

Thirdly, some form of organization is needed to ensure that responsibilities are appropriately assigned, work is carried out on schedule and money is wisely used. In addition, decisions have to be taken and implemented at the right time, and problems are solved when they arise.

Lastly, in every activity there is a need to assess the current demand and to be able to sell the appropriate products or services to customers. It is also necessary to anticipate future demand by moving into other areas of opportunity and to continue development work where necessary.

For easy recall, we can summarize the fundamental elements of business activity as being represented by four Ms:

- Market
- Management
- Money
- Manpower

The last item represents all human resources.

This is illustrated in the sketch in Fig. 2.1, which shows the basis of a business activity as a four-legged stool with each leg representing one of the four elements. It should be no surprise to anyone with a knowledge of statics if we suggest that this business stool is most stable when the four legs are equal. In practice, it may not be possible to achieve this degree of equality at all times, but every effort should at least be made to achieve a balance. For example, a business with plenty of funds but which is very slow in building up its customer base and a team of staff with the required expertise and management skills would be equivalent to having a stool with one long leg and three short legs: not the most desirable platform from which to run an organization!

To achieve success, particularly in the longer term, an organization must get to know its markets, possess high-quality manpower, achieve efficient management, and have access to adequate money. Furthermore, this should be done in a balanced way and, if necessary, in small incremental steps.

2.3 THE FOUR ELEMENTS

The four Ms are regarded as the fundamental elements of a business activity, and each in turn is made up of a number of components or topics. It is quite difficult to differentiate clearly which components belong to each of the Ms because they are all closely

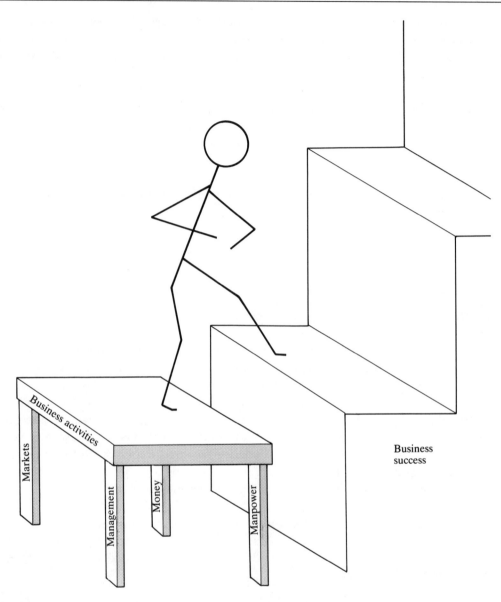

Figure 2.1   The four Ms stool for achieving business success

interrelated. However, it is useful to employ this concept so that a basic framework can be established for acquiring an appreciation of the subject. Before proceeding to a more detailed examination it will be helpful to outline the components making up each of the Ms.

Market

The components of 'Market' are concerned with the requirements of customers, and involve the following:

- Grasping the principles of marketing and their application
- Doing market research
- Identifying opportunities
- Choosing the correct time to take up opportunities
- Establishing the quality of the product or service
- Identifying competitors and the level of competition offered
- Learning selling techniques
- Advertising and promoting
- Providing attractive products and services
- Carrying out public relations activities
- Introducing inventions
- Taking up various kinds of innovation

Management

'Management' deals with coordination of resources to meet needs, and its key components are as follows:

- Defining objectives and devising strategies
- Directing policies and decision-making
- Understanding the roles of specifications and contracts
- Appreciating legal implications
- Undertaking negotiations
- Planning work programmes to ensure schedules are met
- Monitoring progress
- Establishing the quality of products or services
- Making effective use of information
- Encouraging research to overcome specific problems
- Developing new products or services

Money

The term 'Money' covers both sources of funds and methods of judging the financial viability of business activities. For convenience it also embraces government policy, since this can have an indirect but often important effect on business activity. The components of interest here are:

- Understanding economics and its implications
- Identifying sources of funds
- Borrowing capital resources
- Devising investment strategies
- Budgeting of resources
- Comprehending accounting methods
- Determining project viability
- Interpreting government policies
- Calculating the investment requirements of new technologies
- Estimating the cost of projects
- Making full use of available financial resources
- Assessing financial performance
- Taking account of tax rules

Manpower

This term covers both the personal qualities and skills of those needed for the effective performance of a business activity and also the ways in which its human resources can contribute most profitably to an activity. The following components are involved in 'Manpower':

- Recognizing the human qualities required
- Acquiring professional competence in a given subject
- Developing communication skills
- Recruiting and retaining quality staff
- Working as a team
- Taking on a leadership role
- Organizing one's time effectively
- Encouraging continued self-development
- Devising training programmes
- Appraising performance
- Writing job specifications
- Organizing people for efficient performance
- Recognizing the contributions of psychology to work motivation
- Harmonizing working relationships

Clearly, it is neither possible nor desirable to examine every one of these components in this book. Six significant aspects have been selected for each M element, and these will be treated in the next four chapters. The aim is to provide the reader with some fundamentals and understanding of business activities. To overcome any tendency to see each M in isolation, however, several examples of business situations will now be used to illustrate the importance of achieving a balance of all four.

2.4 EXAMPLES

To support this 4M analogy, four illustrative examples have been selected: major companies in Europe; research groups typical of those found in most parts of the world; small companies in the UK; and college students in North America.

Major companies in Europe

Organizations that perform well are usually found to have the four Ms in balance, while those that do less well have failed to achieve this balance within a reasonable period of time.

One car manufacturer in Germany is producing a very successful series of high performance vehicles, ranging from family saloons to super-luxury models. It identified a market sector in which to specialize, and was able to obtain financial support for this. The staff are efficient and the company organization is first-class. The company is fulfilling the requirements of each of the four Ms to a significant degree, and doing this is a balanced fashion. Not surprisingly, it has been one of the most successful companies of the 1980s and will probably continue to be so in the 1990s as well.

Another major European company, in the electrical and electronics industry, has frequently been criticized for its lacklustre performance during the 1980s. Closer examination shows that it has plenty of money (something often referred to as a 'cash mountain'), a staff and workforce of high quality, and an efficient management team. The problem seems to be that it does not know the areas of business on which it should concentrate in the long-term—in other words it is weak in marketing. Suggestions as to what the company should do with its cash mountain range from spending a significant amount on innovations and high-risk projects to giving the money to the shareholders. Someone even advocated a once-in-a-lifetime bonus payment for the 'workers'!

A typical research group

Researchers in almost any country who are building up a group to undertake a project commonly find that they cannot achieve a balance of the four Ms.

It may be that they have identified an interesting project with potential, persuaded staff to work on it, and are capable organizers, but are having difficulty in obtaining sufficient funds to enable them to meet their objective. On the other hand, they may have the funds, a topic of outstanding relevance, and organizational support but are unable to find suitably qualified research staff.

In other situations the principal researcher may have identified an attractive topic of major practical relevance and been given funds to do the research, but lacks understanding of the organizational effort needed to ensure successful completion of such a project. Typical examples include a failure to appreciate the importance of producing results on time and of keeping to the agreed work programme. This difficulty may also be coupled with problems in recruiting and retaining staff with the required expertise to deal with all aspects of the project. Here the difficulties are related to the 'lengths' of the 'Management' and 'Manpower' legs, and unless they are solved the chances of getting an extension of time or new funds are very low!

Small new companies in the UK

Statistics have shown that over 75 per cent of small new companies fail in their first year of operation. Studies have identified a variety of reasons for this, but they all point to an imbalance in the four Ms. In a typical example, the person behind the venture is usually very knowledgeable about the product or service in question, wishes to make a success of the venture, has obtained financial backing for it and has persuaded suitably qualified people to join the team.

The problem is often that the initiator works almost too hard and lacks the ability to delegate. This imbalance in the 'Management' leg of the business stool inevitably results in organizational chaos. In other cases, the failure is caused by under-capitalization. There is a generally held belief that a project which does not need a lot of money is more attractive to potential funding organizations, and that a super idea can generate profit fast enough to overcome any shortfall. Both views can result in inadequacy in the 'Money' leg.

Self-funding college students

Full-time study for an additional qualification can also be regarded as a business activity, and this is particularly true for students in North America, where funds have to be found to pay for a college education. For the mature student returning to full-time study after some years of work, however, success depends on also achieving a balance of the four Ms. Generally, this type of student knows what qualification is needed to improve his or her career prospects, is highly motivated, and has the background skill and experience to cope with the demands of the course. Such a student would, however, be unlikely to achieve the objective within the time planned, unless financial worries are overcome and life is organized so that other commitments are receiving proper attention. In other words, the 'Money' and 'Management' legs need to be balanced with the other two.

In many cases, however, it is the 'Money' leg that causes problems. Typical examples of these include:

- Falling behind in studies because of part-time working to supplement the living allowance.
- An unexpected drop in the real value of savings, caused by an escalating rate of inflation.
- Failure to estimate fully the 'hidden' costs, such as that for transport.
- Inability to adjust from a comfortable lifestyle to that of a 'poor' student.

2.5 BALANCED DEVELOPMENT

It would be helpful to consider how best to ensure that the engineer develops an understanding of the four elements of business in a balanced manner. To a large extent this will depend on the individual's practical needs, job(s) and personal interests.

An individual's objective may be to gain an all-round appreciation of business so as to improve interaction with non-technical personnel and incorporate commercial considerations into decision making. In this case the study of the four elements should concentrate on the fundamental principles of business and their application in practice. If, however, the objective is to become more fully involved in business activities or further studies it would be useful to identify the most suitable approach and to adopt an effective learning strategy. The majority of likely approaches can be classified under the two headings of 'systematic' and 'random' approaches.

The systematic approach assumes that different aspects of business skills and knowledge are required at different stages of an engineer's career. Forward planning is therefore done to identify the needs for each stage and to provide training in good time. For example, the career pattern of one engineer may consist of the following phases:

- Studying at a higher educational institute
- Working as a graduate engineer
- Taking on a project manager's responsibilities
- Holding a senior management position

To achieve a balanced development of knowledge and expertise in this case, the emphasis on the four elements of business would be different during each phase. A student, for example, would benefit most from developing ability in time management and an understanding of marketing. A technical director will want to concentrate on leadership skills and financial expertise.

The random approach, on the other hand, assumes that little formal education in business topics has taken place, but the time comes when the engineer moves from a mainly technical position to one where a greater emphasis must be placed on commercial affairs. The move may be, for example, from design engineer to project manager. In this case a very steep learning curve may have to be climbed quickly and there may even be an over-emphasis, in the short term, on one element of business to the exclusion of the other three. Special attention should therefore be paid to ensuring that the fundamental principles are fully grasped before appropriate components of the elements are selected for further study. The choice made at this latter stage should combine provision for long term needs with what is required to cope in the immediate short term—not forgetting that a balanced development of the four Ms is essential!

These approaches will be discussed more fully in Chapter 8, once the four elements of business have been examined in greater detail in the next four chapters.

CHAPTER 3

MARKETS

This chapter comprises seven sections. Six topics have been selected from the general subject of 'Markets' and each is considered under the headings of: Goal, About the subject, Key questions and issues, and Illustrative examples. The chapter then ends with a list of materials for further study.

Marketing is examined first, with a statement of its fundamental principle, factors influencing markets, and a study of how the best use can be made of marketing. There follows a section dealing with *Market research*, outlining the basic forecasting techniques and highlighting their usefulness and limitations. Methods of carrying out market research are also discussed. The next section considers the term *Opportunities*, and their identification, evaluation and prioritization are briefly examined. The importance of timing is stressed and its influence on business success is indicated by means of illustrative examples.

The section on *Quality* begins with a definition of the term and goes on to discuss responsibility for maintaining quality. Its importance in relation to competitiveness is also highlighted. The section on *Selling* employs a broader definition than simply 'exchanging a product or service for money'. Rather, it encompasses the effort necessary to persuade an individual, an organization or the public at large to accept ideas, as well as products or services, in exchange for some benefit to the giver. This could be money, a positive response, agreement or general support. The aspects considered in this section include the techniques of selling and factors which affect success. In order to make progress an organization has to introduce various innovations from time to time, and, where appropriate, to bring out new inventions. Definitions of these two terms are followed by a brief discussion of their conception, initiation and practical implementation.

3.1 BASICS OF MARKETING

Goal

To ensure that an organization has given full consideration to the needs of potential users in the products or services it provides.

About the subject

What is 'marketing'? A study of marketing textbooks would yield over fifty definitions of the term. The majority of engineers see it as either 'selling' or 'generating consumer preference for' a given product or service. This is hardly surprising, since it is only in recent times that applications of marketing techniques to consumer supply have made the term a familiar one, and few people appreciate that it is actually a well-established discipline. It achieved credibility with managers when it was shown that marketing techniques were able to identify customer preference within a range of competing products or services, something that is particularly useful when the total range available could lead to supply in excess of actual demand. As a result, however, managers tend to be more familiar with the 'new' usage of the term.

It is unlikely that a single, fully comprehensive, definition can be found, but any organization's marketing activity must include the following:

- Identifying the needs of existing and/or potential customers
- Quantifying these needs in terms of numbers of a given product or a service with clearly defined specifications
- Forecasting when these needs will arise
- Examining the company's objectives and strategies in the light of the information gathered
- Determining what percentage of the total market could be won over a given period, i.e. the size of the achievable market share
- Communicating the resulting policy to everyone in the organization
- Determining the resources and facilities required to ensure that the identified needs can be satisfactorily fulfilled

'Satisfactory fulfilment' means profit for the supplier and enjoyment of a quality service by the customer.

Key questions and issues

The fundamental principle of marketing

Like other established disciplines, marketing has its own theories and practices. An understanding of its fundamental principle does much to develop engineers' awareness of marketing 'thinking', and makes it easier to apply these principles in the context of their own work.

The fundamental principle of marketing can be stated as follows:

Supply is conditioned by demand

Alternatively, if there is no demand, there should be no supply. The validity of this principle is obvious, since it is always easier to sell something that customers actually require than to persuade them to buy unwanted goods! There may be plenty of sound reasons for a lack of demand, but acceptance of the principle will do much to ensure that the effort is in the right direction.

Key factors relating to the principle

Closely related to the fundamental principle are the following factors:

- Marketing begins with market research
- Market research should be continuous
- Marketing should have a key role in strategic thinking

The starting point of any marketing exercise should be the gathering and analysing of information on relevant needs in order to identify 'real' business opportunities. This activity must be a continuous process for two principal reasons:

- Needs may well alter almost as soon as information is gathered and analysed
- To be able to respond positively to customer needs an organization must have access to up-to-date and properly interpreted market information

Market research may not provide all the answers, but it allows variations in demand to be identified and taken into account as they occur. The findings will then help to guide suppliers' planning and decision making in the most effective direction.

Factors influencing the markets

A grasp of the fundamental principle of marketing and the implementation of the related factors is essential, but will not necessarily enable a supplier to identify needs in quantitative terms. The difficulty here stems from the number of variables which can influence the market. The situation is not helped by the added complication of having to take 'subjective' elements into consideration. A sudden change in customer preference, for example, may be based on personal preference at the time.

Key factors affecting markets include (see Fig. 3.1):

- The level of world trade
- Politics and government policies
- Technological developments
- The cost of energy
- The degree of competition
- Special situations

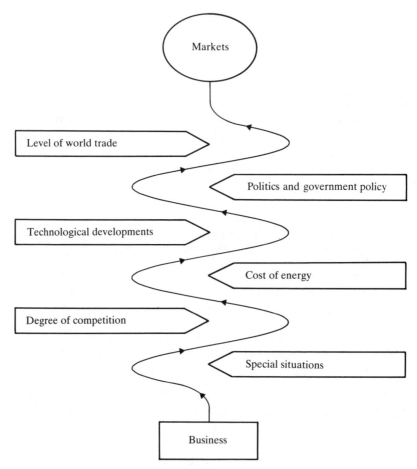

Figure 3.1   The factors that influence markets

The level of world trade

By far the most important factor influencing the activities of a business is the level of world trade. World trade is the process whereby different nations buy goods and services from each other, depending on what they have to offer. A typical pattern would be for one country, e.g. Ghana, to sell a raw material such as cocoa to the UK and to use its earnings to purchase manufactured goods.

World trade is important because the vast range of goods and services in demand these days makes it unlikely that any one country could be totally self-sufficient, or would find it economic to produce all its own requirements. This has resulted in a situation of active trading between nations, and hence organizations are finding many opportunities to 'do business' with companies in other countries. Should a recession occur, however, the demand could very speedily decrease, and even the best-planned marketing strategy would cease to be effective.

Politics and government policy

Organizations and their products and services can be affected both favourably and adversely by government policy. Typical examples of relevant factors include:

- *The level of taxation*: Reductions in taxation can stimulate a consumer demand for luxury goods such as electronic equipment. They also provide an incentive for undertaking projects such as the production of oil from less economic offshore reservoirs. Correspondingly, an increase in taxation can make many business activities unprofitable.
- A *change of government*: Major effects on marketing policies can result from a change of government, whether through an election or the departure of a key minister, or by means of a *coup d'état*.

Politics also affect markets when some countries are closed to a particular type of goods. For example, some advanced computer systems are on the USA's restricted export list and cannot be sold to certain countries.

Technological developments

Technological advances and breakthroughs can be expected to open up fresh markets to organizations alert to opportunities and able to undertake development. Typical examples of this would include:

- *Composite materials*: The successul development of glass reinforced fibre and other materials with special properties has provided manufacturers with alternatives to metals and alloys which have already brought about profound changes in products.
- *Transportation*: By volume, marine transportation carries about 95 per cent of the

goods involved in world trade and by value it carries 80 per cent. The cost and speed of ocean transportation have been considerably affected by the replacement of conventional cargo vessels by container ships. Standardization and the establishment of an effective infrastructure have reduced the cost of transportation, and the increased frequency of sailings allow goods to be delivered in more quickly.

- *Robotic production*:   Many repetitive tasks in manufacturing are being taken over by robots. The majority of car manufacturers have now adopted this form of production where possible, which has both brought down unit cost and increased their ability to meet customer demands.

Cost of energy

The functioning of any business organization depends heavily on the availability of energy, and this will affect the cost of the products or services it offers. It is essential for suppliers to have access to forms of energy which are not only safe and inexpensive but also environmentally friendly. Markets are affected by fluctuations in the cost of energy. A typical example of this is the price of oil in relation to electricity and transport:

- *Electricity*:   Electricity can be generated using a variety of fuels, but at present the industry depends mainly on oil, backed up by coal, natural gas or nuclear fuel.
- *Transport*:   Motor vehicles are the major method of moving goods to customers overland.

In both cases the cost fluctuates with the price of a barrel of oil.

Degree of competition

Current and potential competition and the ease with which market entry can be achieved all affect the market viability of an organization's product and services. Being the sole supplier of a product gives an organization temporary dominance over the market, but as other suppliers recognize the opportunity and provide similar products its share can expect to be reduced. Two examples of this principle are as follows:

- *Transportion*:   The freight rate or cost of transporting a tonne of oil from the Persian Gulf to Europe will depend on the number of tankers available to carry the cargo. When more tankers are bidding for a limited amount of cargo this brings down the freight rate, and vice versa.
- *Car manufacturers*:   With more and more manufacturers offering a similar range of cars in a given price range, it is clear that in order to maintain their market share each will have to bring down prices or else provide more 'extras' while maintaining the same price level.

Special situations

There are many other factors which can affect markets, including:

- *Finance*: Almost every business activity uses borrowed money and is thus affected by rates of interest. Consequently, payment arrangements can vary widely. They range from a 10 per cent deposit when the contract is signed with further instalments depending on progress, to using money advantageously borrowed for a 10-year period without any initial deposit being paid.

- *Past experience*: Customers are very much influenced by their own previous experience and that of contacts. Good previous experience ensures that the loyalty of customers is retained while an unpleasant experience will have the opposite effect.

- *Attitude*: Customers' perceptions about products play a key part in their choice. If, for example, a product is believed to be harmful to the environment, its appeal to potential customers could diminish very quickly.

How can the best use be made of marketing?

In view of the uncertainties involved, many engineers consider that a 'hunch' may be as good as any formal marketing effort, and that decisions should be based on 'common sense'. There are, however, two good reasons why those involved in technology need to give serious attention to marketing.

The first is that it helps to differentiate between good ideas with commercial prospects and those that are simply technological 'inspirations'. Too often in the past engineers have come up with ideas that are technologically brilliant and have been able to make a strong case for their introduction to the market, without taking the precaution of assessing customer need for such provision. The result has sometimes been financial disaster.

The second point is that a grasp of the fundamental principle of marketing and its related factors helps engineers to develop greater flexibility in their approach to change. It could enable them to turn a lack of demand for one particular product or service into an opportunity to develop fresh products and services. Instead of maintaining an unreasonable defence of the current status they will be encouraged to think more positively at all times.

ILLUSTRATIVE EXAMPLES

Example	**3.1(a)**
Subject	**Approaches to marketing by one entrepreneur**
Background	In general, people wishing to start a business do so on the basis of an activity in which they are either interested or experienced, or both. Often no proper marketing policy is defined and applied before proceeding with the venture. Yet the likelihood of success is greatly increased by application of the basic principles of marketing.

Take the case of a Canadian engineer who returned to Vancouver in 1989 after gaining his Ph.D. at a UK institution for work on advanced computing techniques associated with offshore work. At that time, the offshore and subsea market on Canada's west coast was in a depressed state, and he decided to start a business based on his own past experience. He set up a delivery service for luxury yachts and motor craft between Vancouver and locations along the west coast of both Canada and the USA. Unfortunately this venture was wound up after only four months due to poor business prospects over the longer term.

Having elected to make another attempt at setting up a business the engineer decided to look over the notes on a course entitled 'Marine Enterprise' which was part of the study programme he had followed. He paid particular attention to the section on marketing, and noted its stress on the importance of identifying the needs of potential customers. He had already realized that activities involving yachts and small pleasure craft were very popular in the Vancouver area. A short study revealed a high incidence of damage to boats through grounding, collisions with floating timber, and so on. Many owners of these leisure boats had complaints about both the high cost of repairs and the time taken to have them done. He thus identified a clear business opportunity for someone able to provide a speedy and low-cost repair service to these people. It was not long before he had set up such a service, offering an average turn-round time 25 per cent faster than his competitors and at prices below theirs. During the first year he built up a small team of skilled technical staff and by its end this new company had picked up a significant share of the local market.

Comments   This engineer began his first venture without any appreciation of the actual level of business available in yacht and small craft delivery. He thought there was a market in this sector, but soon found his mistake. Once he had identified a clearly defined customer need, however, his determination and flair soon enabled him to develop a strong and efficient business. He is unlikely to do any research now until he has made his fortune!

Source   Private communication.

Example   **3.1(b)**

Subject   **Marketing ships**

Background   Some of the world's best-known shipowners made their fortunes by operating 'Liberty Class' cargo ships which were constructed during the

Second World War and were not expected to last for very long. However, it was nearly twenty years later, in the mid-1960s, that the demand for replacements began. For a medium-sized shipyard to compete successfully for a replacement order it had to offer a design that would meet the owner's requirements precisely. Yet each owner had his own idea of the ship characteristics that would provide the competitive edge. Market analysis indicated that profitable operation depended on how the ships were used rather than on their individual characteristics. The owner's profit would therefore be increased if both capital and operating costs could be reduced.

The 'obvious' solution for one shipbuilder was to build standard vessels with the same equipment so as to simplify crew training and increase the efficiency of maintenance.

The standard vessel design actually evolved, known as SD14, was one of 14000 tonnes with a speed of 14 knots. Over 180 were sold during the next twelve years.

Comments This market opportunity existed at a time when a 'standard' ship proved to be the solution for bringing down production and operating costs. The shipbuilders Austin and Pickersgill carried out an offshore market analysis before embarking on the project. The biggest challenge was to persuade the first customer to try out the new design. Once that had proved successful others soon followed.

Source Conference seminar discussion.

Example **3.1(c)**

Subject **Marketing studies**

Background Sound marketing studies can only be done if the terms of reference are clearly defined. Two examples are given here to illustrate this point.

- *An unrealistic goal*: Some years ago a major US car manufacturer initiated a study to identify the characteristics of the public's 'ideal' car. After two years' work, however, it was clear that such a car could not be specified. Indeed, a car incorporating all the required features would in the end be far less 'ideal' than the majority of traditional designs already on the market. The question posed was really an impossible one and the answer would have been quite ridiculous. It is therefore not surprising that a universally acceptable set of characteristics could not be found.
- *The misuse of marketing studies*: One small company which offers to do market studies won a number of contracts in the UK and Western

Europe. These contracts were to provide market information to assist the management of the client organizations in formulating investment policies. A careful examination of some reports and follow-up of this company's recommendations reveals several common features, e.g. the reports were highly readable, but lacked substance and any real understanding of the key problems; the data provided were often incorrect or out of date; the recommendations made showed little imagination; and the fees charged were very high. That such a situation can continue may seem surprising, but the explanation is straightforward. The managements of the client organizations were simply using these market studies to justify their (often unsound) investment decisions. What better response to potential criticism than the claim that the decision was reached after careful consideration, using the 'best' market information available at the time and from a 'recognized' company?

Comments   To develop a successful investment policy a business organization must use market information that has been competently and objectively gathered, and is unaffected by preconceived ideas.

Source   Private communications and press comments.

3.2 MARKET RESEARCH

Goal

To provide quantified information on the probable demand for one or more products or services within a forthcoming period of time.

About the subject

Market research is one of the tools of an organization's forward planning, being the gathering and analysing of information in order to forecast future opportunities, particularly those with commercial potential.

It must be recognized that the future is an unknown quantity. The task of forecasting is to predict that the position at a given date will approximate to Situation A rather than Situations B or C. Thus, the basic limitations of market research are that:

- It can do no more than narrow down the area of uncertainty within a decision framework.
- Its results are valid only so long as the assumptions behind the forecast are correct and unchanged.

The results of market research are therefore dependent on time. It is, however, an important exercise, since good market research helps to promote effective decisions on policies regarding an organization's direction of interest, commitment of resources and types of product or service to be developed. Likewise, inaccurate results may lead to unwise policies and consequent problems.

There are many market research techniques and a key feature of good research is the use of several different methods in order to cross-check results. The principal approaches are:

- Analysing available information to identify trends. This may include 'desk research', e.g. to find a link between demand and factors such as the economic performance of a nation, or the analysis of internal sales data.
- Using questionnaires to seek relevant information. Varying sample sizes will be used depending on requirements.
- Coordinating views obtained from personal contacts with the results of experience in order to quantify demand.
- Adopting a combination of the above approaches.

Accurate market research depends on the degree of objectivity achieved during the study.

Key questions and issues

The principal approach

The classical approach to commercial forecasting is to collect and analyse information about past and present activity by various means such as questionnaires. This is then used to establish future trends by coordinating the data and plotting graphs of specific variables against time. These techniques are most effective when only small changes take place and when there are few external factors likely to affect the situation. They cannot be expected to yield accurate predictions if, for example, fresh competitors enter the market or if demand is fluctuating wildly. The development of specialized computer software, however, means that much more detailed information can be analysed and patterns in the variation of demand can be identified. Forecasting techniques have thus become more useful for indicating 'firm trends'.

The use of questionnnaires

The questionnaire is regarded as a particularly effective method of gathering both factual information and opinions. Essentially, a set of carefully worded questions is put to a selected sample of people and the replies are then analysed using statistical methods. From the results, qualitative conclusions can be drawn and trends established.

The success of a questionnaire depends on two sets of factors. The first set comprises the design and the wording of the questions. Special attention should be given to the layout so that the questionnaire is uncluttered, and a format should be selected that will help the respondent. The questions themselves should not be too general, but sufficiently specific without being too restrictive. This is vital, as useful conclusions are unlikely to be drawn from the response to vague or ambiguous questions.

The second set of factors is concerned with the size of population sample used and the percentage of respondents, both of which can affect the usefulness of a questionnaire. Market research on a typical engineering product involves circulating questionnaires regularly to between 500 and 1500 people, while the response rate is normally between 4 and 12 per cent.

Forms of questions in a questionnaire

Four basic forms of question are used in a questionnaire in order to obtain the views of those responding. These are:

- *Direct choice*: These questions would provide clear data on preference for one type of product rather than another.
- *Prioritization*: Information on customer's order of priority for various factors is useful, although the compilation of the necessary questions requires careful thought.
- *Rating*: Customers' ratings of different design features are a helpful guide in the development or modification of a product.
- *Comments*: A questionnaire would be incomplete without giving responders an opportunity to make comments of their own.

The role of customer contact

The personal interview has a certain value in market research. It provides the opportunity for customers to comment on the current situation and likely developments, and helps to verify data gathered by other means. It is particularly valuable as an input to forward planning in areas where a specialized knowledge of particular needs may prove crucial. However, it can also have serious pitfalls. Data obtained by this means can be very subjective if the individual's opinion is a personal 'hobby-horse'. Also, it often has only a short lifespan. Thus, unless cross-referencing is done customers' contributions could actually be counter-productive.

A possible procedure

As indicated earlier there is a range of market research techniques and to a large extent it is the product or service in question that determines the choice of approach in a given study. However, it is possible to generalize the key steps to identify the key elements of a market research procedure. The one proposed consists of the following six steps (see Table 3.1):

Table 3.1 A market research procedure

STEP	ACTION	TASK
1	Examine	background data, basic relationships and demand trends
2	Acquire	relevant information about markets and customers
3	Identify	important features attractive to customers
4	Quantify	the requirements and how they are to be met
5	Assess	the strengths and weaknesses of competition
6	Predict	the likely future demand

- *Examine*: Study the background information and relationships between supply and demand as they affect the item in question and, from the information available, establish the trend in demand.
- *Acquire*: Acquire relevant information about the market, the factors with the major influence on its behaviour and existing and potential customers. This will lead to a clear picture of their objectives and their modes of operation.
- *Identify*: Seek to identify the features of the product or service that would be critical in attracting customers.
- *Quantify*: Quantify customer requirements (such as speed of delivery) and price constraints and establish to what extent these can be met. The relationship between demand and price must also be considered in order to determine a target price that should make the new product or service attractive to potential customers.
- *Assess*: At this stage the strong and weak features of the organization's competitors, both existing and potential, should be assessed. This is also the time to decide on the most appropriate modes of communication with customers and the best selling techniques for the situation in question.
- *Predict*: In order to gauge realistically the organization's share of the possible market it is necessary to predict the potential demand. This is done by determining the present size of the market and estimating the scope of untouched areas in order to assess the total size of the future market.

Supporting techniques

Closely related to market research is a powerful technique known as 'scenario planning'. The researcher postulates a range of situations and, having considered their various implications, outlines the best response to each set of circumstances. For example, when the price of oil stood at $30 a barrel, scenario planning techniques were used to examine the possible impact if, at some date in the not-too-distant future, the price per barrel were to rise as high as $100 or plunge to $8. Such forward thinking allows an organization to respond quickly to changes in the market situation and to take full advantage of whatever opportunities it may offer.

Pitfalls in market research

Because the future is unknown, there are pitfalls to be avoided in market research. The most likely dangers are:

- *Asking the wrong questions*:   The investigation is focused on non-crucial issues, which means that future decisions will be based on inadequate and unbalanced information.
- *Inappropriate use*:   There is an unhealthy tendency to use selected market research results to support personal wishes on policy. A study may be commissioned, for example, to provide support for a policy decision already planned by an organization's chief executive. The way in which its findings are applied may well give a market picture with little resemblance to reality.
- *Poor timing*:   Since market research results are time-dependent, snags can be expected if implementation of recommendations is delayed. Conversely, action could be taken too early when results suggest that there will be no market opportunities for many years ahead.

ILLUSTRATIVE EXAMPLES

Example   **3.2(a)**

Subject   **Doing market research**

Background   Volkswagen had been selling cars very successfully in the US market for many years. A market study forecast that profit would be increased if the company moved its manufacturing plant for one particular model to the USA. The main advantages this offered would be a reduction in total production costs and a way of overcoming a possible limit by the US Government on foreign car imports.

When the proposal was implemented, the sales of Volkswagen cars showed a sharp decline. The reason for this was that German products had a high-quality image in the USA and Volkswagen cars were bought for prestige reasons. This factor had been lost with the cars being manufactured in the USA.

However, application of the suggested six-step procedure for market research would probably have had the following outcome:
Step 1   An examination of trends would indicate a good demand for Volkswagen cars and the relationship between price and demand.

Step 2 Acquisition of relevant information would provide the key reasons for the demand.

Step 3 Identification of the features important to customers would yield the crucial significance of manufacture in Germany as a prestige factor.

Step 4 Quantification of other customer needs would suggest that the requirements could not be met by transferring manufacture to the USA.

Step 5 A gauging of the strengths and weaknesses of competitors would yield the factors with the most crucial role.

Step 6 Realistic predictions could therefore be made about the appeal of an American manufactured Volkswagen.

Comments It would seem that the market research study actually carried out placed a great deal of emphasis on production costs and legal obstacles and failed to identify the prestige factor of German manufacture. A systematic market study procedure should enable such features to be identified, and allow the development of appropriate strategies for selling the product or service in question.

Source Press comments.

Example **3.2(b)**

Subject **Developing a postgraduate degree course**

Background Educational institutions can use a number of different approaches to the development of courses. Examples of typical starting points are: the transfer of research advances, the decision to make better use of available expertise, the personal interests of the academic staff, or the recognition of a lack or a potential need.

Not many institutions use a full market analysis in determining what courses to offer, but this was the case prior to the development in the late 1980s of a new Master of Engineering course in Manufacturing Systems by the Lawrence Institute of Technology in Michigan, the largest private university in the USA.

Example **3.2(c)**

Subject **A sample questionnaire**

Background The item selected for market research was an electric fan capable of providing either cool or hot air, and the questionnaire was to offer responders four methods of expressing their views on its design. These were as follows: direct choice; prioritization; rating; comments.

Figure 3.2 shows the resulting questionnaire on fan heaters/coolers.

Comments   Better results can be expected from a short questionnaire which is sharply defined than from a long one with too much detail.

Source   Private communication.

The key steps in the approach were as follows:

- A clear 'mission objective' was developed that could win the support of the academic staff.
- An internal assessment was done to identify the strengths and weaknesses of the faculty, the laboratories and the library.
- Agreement was reached that the course would be staffed by seventeen out of the ninety members of the Engineering Faculty plus three members of the Management School staff, all of them having Doctor of Philosophy degrees. Eight laboratories with facilities valued at US$7 million would be used for various aspects of the course.
- A market survey questionnaire was sent to 4346 people in industry and achieved a 12.1 per cent response. Of these, 268 individuals expressed an intention to take the course.
- A survey questionnaire on 57 colleges in the USA offering similar courses obtained a 68.3 per cent response. Industrial personnel indicated that, of the comparable courses already available, eight were of significant value. The support for these ranged from 55 per cent to 75 per cent of the response sample.
- On the basis of the market survey results, the subjects included in the programme at the Lawrence Institute of Technology provided a balance of industrial and academic features, and specifically offered desirable features not available at other institutions.

Comments   The needs of users were taken fully into consideration in the preparation of this higher degree course, and the preliminary survey of comparable programmes ensured that the optimum combination of topics was achieved. This case illustrates how a businesslike approach can be adopted in devising educational programmes. To obtain information of real value, however, it is essential that questionnaires are very carefully constructed.

Source   Taraman, K.S. and Ellis, R.W., Important steps to successfully develop a new Master of Engineering in Manufacturing Systems program. *Proceedings of World Conference on Engineering Education for Advancing Technology,* Sydney, Australia, February 1989.

Compact domestic fan for space heating and cooling

We should be most grateful if you would fill in this questionnaire, which
is aimed at improving the design of our fan.

1. What would be your main requirement?   (Please tick as appropriate)

    ☐ Heating

    ☐ Cooling

2. How would you rate the importance attached to the following factors when
    selecting a fan? (1 = top priority,   2 = second-top, etc.)

    ☐ Purchase price

    ☐ Reliability

    ☐ Efficiency

    ☐ After-sales support

    ☐ Safety features

    ☐ Other features (Please specify) _____

3. How do you rate the following design features?

Essential	Good	Reasonable	Acceptable	Unimportant	Not relevant	Not applicable	
☐	☐	☐	☐	☐	☐	☐	Appearance
☐	☐	☐	☐	☐	☐	☐	Physical dimensions
☐	☐	☐	☐	☐	☐	☐	Range of settings
☐	☐	☐	☐	☐	☐	☐	Noise level

4. Any additional comments? _____

    _____

    _____

Thank you for your cooperation.

Figure 3.2   Sample questionnaires illustrating four methods of seeking market
responses.

3.3 OPPORTUNITIES AND TIMING

Goal

To identify potentially attractive situations and establish the correct time to be commercially involved in them in order to achieve success.

About the subject

What is an opportunity? It could be said that everything in the world offers an opportunity to someone, but a more generally useful definition might be:

> A situation which, if exploited properly, and at the most suitable moment, could yield sucess.

However, some of the terms used in this definition need further explanation.

A given situation can offer a number of business possibilities and some of these will have potential for commercial success, although the amount will depend greatly on the attitude of those involved. For example, poor sales of an apparent necessity, such as shoes, in a heavily populated country may be seen only as a commercial failure by one supplier while another supplier may see the situation as a vast opportunity.

With regard to the question of the 'most suitable' moment, a supplier who seeks to sell a product or service for which the public is not ready may suffer severe loss. Conversely, a supplier who tries to move into an active market may find that it is already saturated with rival versions of the product.

'Success' means different things in different contexts. Even in commerce the yardstick may vary from monetary reward to increased efficiency, improved standards or personal satisfaction, or it can involve a combination of all four.

Key questions and issues

How are opportunities identified?

'Some folk are good at recognizing opportunities while others just have no luck at all!' That comment may have a familiar ring, but—although it contains an element of truth—the idea of 'luck' has very limited justification. In practice, even 'recognized' opportunities will not automatically lead to success unless properly implemented. However, the ability to identify them is an important factor.

This ability comprises a combination of personal attributes coupled with experience. The key attributes are:

- A positive attitude
- Mental alertness
- A capacity for 'lateral thinking'
- The ability to think out all the implications of an idea from initial conception right through to final outcome.

The 'luck' is usually connected with timing. A company perceives a situation with commercial potential at the very time when it is in a position to meet the need. Alternatively, a member of the staff comes up with a novel response to a current problem, and if this is recognized and implemented quickly it provides—or creates—an opportunity which can be actively exploited.

Initial evaluation of opportunities

It is never easy to tell in advance whether an opportunity will lead to genuine business developments and commercial success. The decision on whether or not to take it further depends on a number of factors, and some of these are considered here:

- *Influence*: In any organization certain key individuals have a special influence on decisions made. This influence may stem from a controlling financial interest, years of experience, or simply the ability to put a case over well.
- *Policy*: The organization may be seeking to diversify its activities and is very receptive to fresh ideas.
- *Procedure*: The organization has an established procedure for judging opportunities systematically and taking the most attractive one on to the next stage of assessment.

Criteria for prioritizing opportunities

It is possible to apply a number of general criteria to a set of opportunities, and these will greatly assist prioritization. These criteria are summarized as follows:

- *Feasibility criteria*: Establish whether the opportunity is feasible from a technological point of view, and whether there are problems still requiring solution.
- *Commercial criteria*: Establish whether there is a real market for the identified opportunity, and critically analyse all available market research data.
- *Organizational criteria*: Establish how closely the opportunity fits in with the corporate objectives of the organization.
- *Credibility criteria*: Establish how the organization will manage in detail the development of the opportunity from the start to its commercial implementation. Identify possible hurdles and ways to overcome them.

By placing a proper weighting on each of these four factors, it is possible to examine any set of opportunities objectively and assign a logical commercial priority to each.

A procedure for finding opportunities

There is no formal procedure that would be suitable for every situation, but the following steps are usually involved in the identification of an opportunity:

- *Observe*: Being alert to what is happening in general, and in the commercial and industrial world in particular.
- *Question*: Asking why the present method is used for a process, product or service.

- *Investigate*:   Investigating similar ideas, products, methods or services, and their cost, effectiveness, competitive standing etc. This is necessary to establish whether they can fulfil market requirements.
- *Gather*:   Acquiring all relevant information on the identified opportunity.
- *Evaluate*:   Using relevant criteria to assess the possible gain to be achieved from taking up perceived opportunities, and to identify potential drawbacks.
- *Decide*:   Make a decision on whether to proceed with a new idea, shelve it, pass it on to someone better fitted to implement it, or find an alternative way forward.

These six steps are summarized in Table 3.2. Intelligently implemented, they can help greatly to systematize the identification of opportunities.

Table 3.2   A procedure for opportunity identification

STEP	ACTION	TASK
1	Observe	what is happening generally
2	Question	why present approaches are in use
3	Investigate	similar solutions and their effectiveness
4	Gather	relevant information
5	Evaluate	the potentials and drawbacks
6	Decide	on a prioritized opportunity

Dealing with timing

The difference between good and poor timing is that between a real opportunity and a lost one. It is not to say that instant decisions have to be made in order to demonstrate a 'dynamic response to opportunities'. As indicated earlier, timing is far more a case of recognizing the appropriate moment to introduce a product or a service to the market. Too early, and the item on offer will encounter customer resistance and a long struggle to establish itself. Too late, and other companies will have already consolidated their hold on the market.

There is no magic formula for determining the correct moment to exploit an opportunity, but two factors can help. These are:

- *Signals*:   Customers send out signals in many forms and they must be correctly interpreted. Regular analysis of market research results will contribute greatly to effective timing.

● *Experience*:   Practical experience of business activities is valuable. It is particularly useful to make a critical analysis of the role of timing in both sucdessful and unsuccessful commercial ventures.

ILLUSTRATIVE EXAMPLES

Example   **3.3(a)**

Subject   **Designing cameras for the mass market**

Background   In the 1970s increasingly complex cameras were being produced, and all over the world tourists could be seen labouring under a self-imposed burden of cameras, lenses, filters and stands. Market research began to indicate that the growing demand was not for greater sophistication but for a compact, lightweight, easy-to-use, or 'idiot-proof' camera. It should also offer automatic focusing, press-button winding and rewinding, and automatic flash when natural light is inadequate. The technologies needed to design such cameras had not yet been developed, much less perfected, but this was perceived as a major market opportunity for the 1980s. Research and development were therefore devoted to enhancing the compactness of cameras, together with the following features:

● Automatic focusing that produced all-round sharpness in the pictures
● Automatic film winding after each picture is taken
● Automatic motorized rewinding by pressing a button
● Built-in electronic flash which responds automatically to low light conditions
● Easy loading and unloading of films

By December 1988 there were 42 'standard' cameras on the market, offering a range of facilities at an average price of £79. Japan holds the dominant position with 24 versions, Hong Kong produces eight, Korea five, Taiwan four and the USA one.

Comments   Manually focused cameras sold well before automatic focusing techniques were developed. Over the past ten or twenty years, however, there has been a steady increase in foreign travel by holidaymakers and others wishing to make a visual record of activities and experiences. It is this non-professional market opportunity that has been very successfully taken up by the compact camera industry.

Sources   'The compact camera' in *Which?*, December 1988. Published by the Consumers' Association Ltd, UK. Press comments.

Example	**3.3(b)**
Subject	**Opportunities and timing**
Background	In 1973, following a decision by OPEC (Organization of Petroleum Exporting Countries), the price of a barrel of crude oil increased threefold over a very short period. This led to the so-called 'first oil crisis'. That crisis affected the activities of every country in the world. Some of the major oil producers became very rich, e.g. Saudi Arabia, and other countries that depended on oil for their industrial development suffered, e.g. India.

The industry which suffered most was bulk shipping. Up to the early 1970s bulk shipping operators were ordering ever larger oil tankers, with dead weights (or crude oil capacity) of between 300 000 and 1 000 000 tonnes. The 1973 crisis resulted in a gross oversupply of these vessels and a large number had to be laid up. In addition, very few new orders were placed with shipyards during the following two or three years.

While other types of shipping became popular, there was great reluctance to become involved in the oil transport business. Typically, second-hand tankers of 250 000 tonnes dead weight were changing hands at around £2.5 million, although they might have cost over £18 million to build just a few years earlier.

Tankers, however, are designed for a lifespan of 20 years, after which they must be retired, scrapped, or subjected to a special repair programme to prolong their working life. By the second half of the 1980s more tankers were being retired or scrapped than new ones were being built, and a number of new shipping companies were formed at that time to exploit this opportunity.

Edinburgh Tankers plc is typical of such companies owning and operating ships under the British flag. It raised £13.2 million from the general public by three British Expansion Scheme (BES) Issues in early 1987, 1988 and 1989, and sought a further £5 million for expansion early in 1990. Under this scheme, subscribers can claim back income tax from the UK Government at the highest rate they pay, on condition that the shares are held for a minimum of five years.

The main factors that attracted Edinburgh Tankers plc to exploit this opportunity during the second half of the 1980s can be summarized as follows:

- *Orders for new tankers*: More tankers were being retired or scrapped than new tankers ordered. For example, by the year 1995, 28 per cent of the total number of vessels in the range 60 000 to 100 000 tonnes dead weight would have reached the age of 20, while new vessel building would replace only 11 per cent of the total.

- *Level of seaborne trade*: With a reduction in the supply of tankers to carry the available tonnage, the freight carrying rate would increase if the level of demand either remained constant or increased.
- *The capital value of the vessels*: The value of operational tankers increased during the period November 1986 to December 1989. The value of the three tankers owned by Edinburgh Tankers plc altered as follows:

(a) The vessel *Edinburgh Taller* was acquired in June 1987 for $10.95 million and sold in September 1989 for $20.2 million.
(b) The vessel *Edinburgh Fruid* was acquired in May 1988 for $15.00 million and by April 1990 was worth $26.5 million.
(c) The vessel *Edinburgh Savannah* was acquired in May 1989 for $7.5 million and by April 1990 was valued at $10.5 million.

This venture, of course, involves high risks. There could, for example, be a sudden downturn in world trade, and conflict in the Gulf region could lead to either a steep increase or wildly fluctuating oil prices and associated uncertainties. Nevertheless this is an opportunity with potential.

Comments   Although it may still be too early to judge the longer-term performance of Edinburgh Tankers plc, the company has already demonstrated that it recognized at the right time the new opportunity for operating tankers when the market was picking up in the late 1980s. It was able to raise the capital necessary to take this opportunity up, partly by borrowing from banks and other sources, but partly by taking advantage of a special scheme being offered at the time by the UK Government.

Source   Prospectus published by Edinburgh Tankers plc in 1989.

Example   **3.3(c)**

Subject   **Timing of an engineering initiative**

Background   In the 1970s oil and gas exploitation in the North Sea depended extensively on divers to perform many of the underwater maintenance tasks. Such tasks included, for example, inspection of structural members, pipeline connection and the replacement of equipment components. Diving, however, is a dangerous occupation, and as exploitation went into deeper waters research was needed to overcome fresh problems. In particular there were difficulties with the gas mixtures used by divers for breathing.

In 1979 the need for a diving research centre was recognized and a committee was set up to formulate plans and seek funding for such a facility to be built at Aberdeen in Scotland. After much discussion and very positive results from market research studies, the National Hyperbaric Centre was officially opened in September 1987 at a cost of over £6 million. It was operated as a free-standing company with Seaforth Maritime, an Aberdeen-based offshore service company, and the Scottish Development Agency as shareholders. The Centre has excellent facilities with three diving chambers, medical research wards, meeting rooms and workshops, and it was envisaged that after an initial set of grants it would become self-financing.

Unfortunately there was very little business activity after the Centre opened and its commercial prospects looked bleak. The longest contract undertaken was work valued at £1 million towards a dive to simulated depth of 450 m (1500 ft), performed in November 1988 with the support of the UK Department of Energy. By December 1989 it was operating on a 'care and maintenance' basis with only four staff, and the liquidation of the company was announced in late March 1990.

Comments

This is a classic example of a missed opportunity due to poor timing. The basic idea was excellent, but such a facility should have been provided ten years earlier when the commercial potential was excellent. In the late 1980s, however, there were two factors affecting the Centre's commercial prospects.

The first was that offshore technologies had advanced very rapidly in the previous decade and many installations were designed for maintenance by means of unmanned, remotely operated vehicles. For underwater maintenance work these techniques had become strong rivals to intervention by divers.

Secondly, exploration and exploitation were increasingly directed towards deeper waters, beyond the range in which divers could operate safely.

Source

Private communications and press comments during the period 1989 to 1990.

3.4 QUALITY AND COMPETITION

Goal

To ensure that the standard of an organization's product or service is more attractive to potential customers than those of rival producers and suppliers.

About the subject

What is 'quality'? Unless preceded by an adjective such as 'poor' this word usually implies 'excellence', which is a subjective concept until agreed assessment parameters have been established. In the present context a 'quality' product or service is one that very closely meets a stringent set of agreed specifications. Take, for example, two companies producing a similar item: one achieves a defect rate of 0.02 per cent while the other only manages a rate of 0.1 per cent. The first company's product would then be classed as one of higher quality, although both organizations may be regarded as producing quality products.

'Quality' also signifies meeting customers' requirements, and the concept provides members of an organization carrying out different functions with a common reference language when seeking to improve a product or service.

What is 'competition'? This is a topic that has been receiving considerable attention, particularly in relation to consumer goods. The term is used here to mean the contest for customer preference between rival companies offering similar products, services or functions. Typical competitors would be two car manufacturers targeting the same range of customers with vehicles of similar engine capacity and price, but differing in, for example, shape, engine performance and internal arrangements. 'Competition' has been linked with 'quality' here because many engineering activities have a direct bearing on the competitiveness of a product through their impact on its overall cost. This is best illustrated by the experience of the Sony Corporation of Japan. In the book *Made in Japan*, Akio Morita, the co-founder and Chairman of Sony, attributes Japan's success in the world market to 'good old-fashioned competition'. He believes that it is in competing for a share of the domestic market that companies develop the capacity to succeed internationally. Consumers become very demanding where there is strong local competition, with the result that suppliers are forced to increase the quality and variety of their products. The companies that survive here do so by critically examining their operations and developing more efficient approaches, which will eventually assist them in winning a significant share of the international market. Sony's own success is due to an appreciation of the importance of competition and to the quality of its products and back-up service.

A joint consideration of the two topics should provide some insight into how this connection operates to the advantage of a business organization.

Key questions and issues

Selecting a quality level

In practice, there is no standardized method for deciding on the level of quality to be aimed at in a proposed product or service. In general, the quality offered must be closely related to value for money. The right decision about quality depends on the results of market research, experience and an understanding of the competition to be faced. It should be noted, however, that quality is a relative phenomenon and adjustments by a competitor can very quickly alter the status of a product or service. Ultimately, increased value for money must be the constant aim.

Does quality cost more?

There is a general belief that quality products or services will always be more expensive. This is not always true, particularly if the basic factors contributing to quality are built into the system through which the product is made or the service provided. Costs can often be reduced through improvements in design, such as simplification, which can lead to increased production efficiency. One type of personal computer is a case in point: when the number of parts was reduced by half and the number of suppliers of components was reduced by a similar ratio, very significant cost reductions were achieved.

Achieving a high level of quality will help to:

- Reduce the incidence of defects
- Minimize maintenance and repair costs
- Cut down inconvenience to customers
- Enhance the reputation of the organization
- Increase the likelihood of obtaining future orders

These are all factors to be taken into account in calculating the total cost. It should be noted also that a defective product may lead to an accident in certain circumstances. The penalties associated with 'product liability' can be prohibitive!

Quality assurance and quality control

There are two terms associated with quality. The first is 'quality assurance' or 'QA', the term used for the approach adopted to ensure that the supplier of a product or service will perform to specification. The second term is 'quality control' or 'QC', which is one aspect of applying quality assurance and involves inspection, measuring and checking to ensure that the agreed specification is met.

In practice there is a whole range of agreed quality standards, each requiring satisfaction of a different set of criteria. Depending on the targeted market, the standard adopted for a product may be either national or one set by an international body, such as the International Standards Organization (ISO). In either case, labelling must include the reference number, e.g. ISO 9000, or BS 5750 to show that it meets the requirements of the British Standards Institute. Many of these official standards offer several 'grades' and a typical example of this is to be found in the classification rules of Det Norske Veritas, Norway. The most demanding level of one of its standards for use by the oil industry involves meeting sixteen criteria, including:

- Organization
- Documentation and charges
- Identification, marking, handling, storage and shipping
- Final inspection
- Quality audit

In some cases a lower level may be sufficient, and this involves fulfilling only a selection of the complete set of criteria.

System for achieving and maintaining quality

Mistakes made in one section of any organization can create problems for others and small errors can also be greatly amplified. Thus, ensuring that tasks are carried out correctly the first time and every time has a number of benefits. The key ones are:

- *Achievement of greater efficiency*: Less effort has to be expended on correcting errors, re-doing work and eliminating or at least reducing possible delays.
- *Cost-effectiveness*: Carrying out a procedure once is always less expensive than having to repeat it in order to achieve a specified standard.
- *Enhanced reputation*: It is easier to sell a service or product if the organization is known for the quality of its output, delivery on time and adherence to the agreed price.

To fulfil such requirements calls for clear management directives and an effective control system. The system that is receiving most attention at present is called Total Quality Management or TQM, and this is the one that has found widespread favour in Japan. It can only work, however, if the principle is accepted by everyone in the organization and it is implemented in every activity.

Responsibility for quality

This responsibility lies not just on the 'Quality Assurance Department', or the 'Quality Control Section'. Because of competition it is everyone's task to look after quality. The following four groups, however, have a special responsibility in this regard:

- *Suppliers*: Suppliers need to foster a consciousness of quality in every member of the organization. This can be done by, for example, setting up formal or informal groups responsible for ensuring that the concept of quality is understood, sensible criteria are selected and methods are devised for ensuring that the desired goals are reached.
- *Commercial customers*: Commercial customers ought to *insist* on an agreed level of quality being maintained so that they can provide the same standard in their own work. A casual attitude in this regard will lead to a general lowering of standards which will have adverse long-term implications.
- *Regulatory organizations*: In certain cases regulatory organizations, such as the British Standards Institute, will provide advice on how to satisfy the standards for a given product or procedure.
- *The public*: It is also the duty of the public at large to expect an acceptable level of quality in the products and services it purchases so that the interests of all are served.

Factors affecting competition

Apart from quality, the key factors influencing the selection of a supplier of a particular product or service can be grouped under the following four headings:

- *Technical specifications*: A product or service has been designed to meet the customer's requirements to a very close tolerance. This might be the case with the output and efficiency of an engine, for example.
- *Reliability*: This factor is closely related to quality and is a crucial measure of the performance of the product or service. A high degree of reliability minimizes the problems caused by failure, and a reputation for reliability increases customer confidence.
- *Delivery on time*: All business activities have tight time schedules which must be closely adhered to, as delays can have serious knock-on effects. Failure to deliver a product to the export agent by a given date, for example, could mean that it misses shipment to the overseas customer.
- *Price*: The price of a product or service must represent value for money in absolute terms and also in comparison with what competitors are offering.

Influence of price on competition

There is a general misconception that to be competitive the price of a product or service must be 'low' and that by reducing the price it will become more successful. In fact, it is only possible to provide a particular level of quality in any service or product at a given price and at a given point in time. Genuine technological breakthroughs, improved efficiency or achieving economies of scale, however, can all reduce production costs and thus provide a competitive advantage. Artificial reductions in price will mislead or provide only a short-term competitive advantage, and can be very counter-productive in the longer term. This is evident in two instances.

In the first place, a service which is significantly lower in price than that of competitors is often regarded by customers as inferior compared with 'high-quality' expensive services! In the second, in order to enter a market a supplier may offer a product at a price below cost in the hope of 'eliminating' competition, but unless this price is heavily subsidized it will not be viable beyond a short period.

To achieve the correct balance is a very challenging task!

ILLUSTRATIVE EXAMPLES

Example   3.4(a)

Subject   **Competition and the pricing of catalytic converters**

Background   In a number of countries, such as the USA, Japan, Sweden and Switzerland, there are very strict laws to control emission from car exhausts. Such controls will operate within the countries of the European Community from 1993, when catalytic converters become obligatory on all new cars. Realizing the harmful effects of exhaust emissions, however, and wishing to be seen as environmentally conscious or 'green', many car manufacturers in the UK market offer a catalytic converter as an option. The extra cost of this ranges from £200 for a simple unit used on smaller and less expensive cars to £850 for a unit for more powerful models.

This situation can be changed very readily by the action of a single manufacturer. From January 1990 Volvo Concessionaire, the UK importer and distributor for the Swedish car manufacturer Volvo, has been offering catalytic converters as an option on all of their models and from that date there has been no difference in price between new Volvos with or without catalytic converters. Furthermore, no attempt will be made to recover the cost of this change by a general increase in prices.

This initiative makes Volvo extremely competitive and other car manufacturers will have to think again about catalytic converters. Possible choices are:

- Bring down the price of their own converter option
- Offer the option at no extra cost
- Seek to provide converters of higher quality to justify an additional charge

Comments　　It is clear that a new competitive factor introduced by any one company will force other car manufacturers to alter their policies in order to maintain their own competitiveness.

Source　　Manufacturer's brochure and press comments.

Example　　**3.4(b)**

Subject　　**Quality resulting from competition**

Background　　An example drawn from the consumer electronics industry is used here to demonstrate the close link between the quality and the competitiveness of products.

It is generally recognized that Japan's consumer electronic products such as televisions, video cassette recorders and compact disc players have gained a major share of the world's market. Many studies have been carried out to determine the reasons for this success. In his book *Made in Japan*, Akio Morita, founder of Sony, says that this success is due to the intensity of competition in Japan. Indeed, he goes so far as to claim that the 'life-blood of Japan's industrial engine is good old-fashioned competition'. The domestic competition is so fierce that only the strongest can survive and the consumer, in turn, has also become increasingly demanding. Once companies have passed the 'domestic' test, they are 'seasoned' for international competition.

Competition is also closely linked to quality. 'Quality' here means that the highest specifications have been met in relation to a large number of requirements, including:

- Design and construction of the concept
- Appearance of the product
- Reliability
- Innovative features
- Efficient after-sales service
- Range of choice

These have also to be linked to price and to people's perceptions, thinking, fashions, taste and interests.

Comments　　To compete successfully by means of quality, the management of a company must accept the need to aim for a specific level of market share rather than short-term profits. For example, it may be necessary to purchase an expensive piece of equipment in order to achieve the necessary quality. This will have to be set against the initial profits from the product, but the company will benefit from the investment in the longer term.

Source Morita, A., *Made in Japan*, 3rd edn, pp. 203–25, Fontana/Collins, London, 1990.

Example **3.4(c)**

Subject **Quality and competitive advantage**

Background In 1987 the chairman of the Morrison Construction Group (MCG) in Scotland tried to identify a means whereby his company could gain a significant competitive advantage over other contractors. His approach was to seek to identify the leaders in certain other commercial and industrial sectors and the reasons for their success. The results obtained all pointed to the critical common factor—quality.

In the retailing sector Marks and Spencer plc was the company whose standard of excellence surpassed that of all of its competitors, their profit per employee being three times that of the nearest competitor. MCG's chairman found that the company worked closely with suppliers in order to offer high-quality products and services, and this was combined with a strong managerial commitment and a high level of staff involvement.

In the chemical industry, Shell Chemicals was selected because it had set out to achieve a reduction of 10 per cent or £60 million in expenditure in order to maintain the same level of annual turnover. It had achieved half of this target by means of a quality improvement programme which had the full cooperation of the staff.

In manufacturing it was Japanese industry which had to be highlighted for its consistently high standard and the strong involvement and dedication of its workers.

No comparable company could be found in the construction industry. Typical examples of lack of commitment to quality here included:

- Consistently late completion of contracts
- Considerable wastage of material on site
- Much reworking needed before jobs are accepted
- Many instances of poor finish

These findings led the Morrison Construction Group to adopt a strategy of commitment to quality which would be applied throughout the organization. Notable factors in this strategy were an emphasis on communicating the objective of quality to everyone in the company and a decision to implement a significant increase in training so as to ensure that all were equipped to produce the required results.

Comments   This study illustrates how performance efficiency, cost saving and increased profit can all be achieved by a commitment to quality. The long-term result of such a policy is competitive advantage.

Source   Morrison, F., Quality for competitive advantage, *Business Journal*, North of Scotland Issue 35, May 1989.

3.5 EFFECTIVE SELLING

Goal

To persuade an individual or organization to give monetary or other benefit in exchange for goods or services.

About the subject

A basic characteristic of business activity is the offer by one party to another of a product, service or an idea in exchange for some, usually financial, reward. In other words, one party is selling something to another. 'Selling', however, is a much broader concept than this, and could well be defined as:

> The process whereby one party persuades an individual, an organization or the public at large to accept an idea, a product or a service in exchange for some reward, whether money, a positive response, agreement or general support.

An effective selling technique is one that achieves its objective, and the 'deal' made justifies the effort expended. Hence, in the course of our lives, all of us have done some kind of 'selling', although we may not have recognized the activity as such. Typical examples of non-commercial selling would be a youngster persuading parents to allow him the use of the family car on Saturday evening if he washes it on Sunday, and a candidate for a local government office persuading voters to elect her if she agrees to take action on local problems.

Sometimes the idea of 'effective' selling is misunderstood—intentionally or unintention-ally. Success is seen purely in terms of ensuring that the other party accepts the product or service, even at the cost of any profit margin, or a drop below the cost price. That, however, is *not* effective selling, whatever may be the possible justification for such measures. Truly effective selling is a business skill that has to be developed and refined, like any other.

Key questions and issues

Can selling techniques be taught?

Selling is a skill, which has to be acquired and improved through practice. For various reasons some people are better at selling than others. This may be due to personality, as some people enjoy persuading others to adopt their point of view, or simply through appreciating the importance of selling in business. Even those who are good at selling can improve their skill through formal or informal training. People who are less skilled would need specific training, together with opportunities for practice in order to improve.

A possible selling procedure

Selling situations vary considerably, depending on the product, service or idea on offer and also on the geographical location targeted. A possible procedure that could result in successful selling involves the following six tasks:

- *Define*:   The objective of a selling exercise must be clearly defined before initiating the preparation, as this is the way to channel effort in the most effective direction.
- *Prepare*:   The seller must build up a picture of target customers, detailed enough to ensure that the item or idea on offer will match their requirements, policies and revealed preferences.
- *Provide*:   Publicity material should be provided, aimed at stimulating an awareness of the item or idea for sale and highlighting its strong features. This can be done in a variety of ways, through brochures and advertisements in journals, personal presentations and on-site visits.
- *Establish*:   It is essential to generate confidence in the sales item and this is fostered by establishing the credibility of the person involved in the selling process and his or her organization.
- *Select*:   Customer response is usually related to perception of the situation at a particular time, and so the choice of moment for action can make all the difference between success and failure. Even a child knows that it is best to seek approval for a proposal when parents are in a 'good mood'!
- *Offer*:   Once a sale is likely, or agreed in principle, skilful support should be offered in order to turn the possibility into a firm deal.

Table 3.3 summarizes these six steps.

Table 3.3   A procedure for selling

STEP	ACTION	TASK
1	Define	the objective
2	Prepare	the case for the target customer
3	Provide	information to stimulate awareness
4	Establish	a credible base
5	Select	the correct timing
6	Offer	support to gain a firm deal

The role of psychology

Successful selling involves some understanding of human psychology and a large number of studies have been done on this topic. Issues that should be noted include the following three.

Firstly, most people want to feel that they have 'got a good deal', or obtained value for money, when they buy something. It is essential to ensure that this belief is always justified, although the ways in which it is done will vary with circumstances. Possibilities range from offering a discount in the price or giving 'free extras' to providing a superior service or introducing a flexible payment scheme.

Secondly, some customers respond to the 'hard sell', while others prefer a 'soft' approach. Wherever possible the technique employed should take account of this fact. Sometimes it is necessary to change tactics during the course of selling if this would appear to give a better chance of success.

Lastly, the image of the person doing the selling is important. Again, no single image will appeal to all customers, but an impression of competence, confidence, sincerity and courtesy will do much to win customers over to a deserving product, service or idea.

How much does credibility count?

Because of the high cost involved, the selling time for some engineering products or services may be quite protracted, particularly in the international market. A single sale can involve patient negotiation for anything from a few months to several years. Some companies may decide that this is too long and opt out of this particular market. If they do decide to continue, however, one of the most important tasks is to establish the credibility of the product or service, the selling organization and its representatives. A precise definition of 'credibility' is not easy, but it does involve factors such as quality and price, efficiency and general track record. In a selling context credibility must include:

- *A long-term approach*: Strategies should be employed that indicate a willingness to do business over a long period and decisions should be avoided that will achieve only short-term advantage.
- *Support services*: Many engineering products are sold on the basis of good after-sales service, and one of the quickest ways of losing credibility is to fall down on this aspect once the sale has been completed.
- *Consistency*: In so far as this is possible, a high level of consistency should be maintained in dealing with different customers.

ILLUSTRATIVE EXAMPLES

Example	**3.5(a)**
Subject	**Selling a new product**
Background	The motorcycle market is a highly competitive one. This example will show how Honda won a dominant share of the US market by effective selling, through implementation of a sound marketing strategy.

Once a lightweight, low-cost machine had been developed Honda adopted the following selling approach:

- *Price*: The cost of the machine was 25 per cent of that of the larger US models.
- *Distribution*: To ensure that this new machine had the salesman's exclusive attention, Honda set up its own distribution agencies throughout the United States, achieving a network of 125 within a matter of months.
- *Image*: Up to that time motorcycle users had had a poor 'image' which inhibited expansion in sales. Honda set about changing the public's perception of motorcyclists by means of a major advertising campaign. The aim was to cultivate the idea that the Honda motorcycle was socially appealing and acceptable to the average family. Typical advertisements in leading magazines such as *Time* carried the slogan, 'You meet the nicest people on a Honda!'.
- *After-sales service*: Great emphasis was placed on after-sales service with the offer of generous guarantees and service support, and readily available supplies of spare parts.

With such a positive approach to selling it is not surprising that the sale of Hondas in the USA expanded rapidly and the objective of growth in the number of machines sold was quickly achieved.

Comment   Honda's objective was to increase the number of machines sold. To fulfil
                 this objective the company identified a niche in the market, worked out a
                 careful strategy and devoted considerable effort to selling the idea of the
                 motorcycle to a broad spectrum of society. The result of this approach
                 was a considerable increase in sales.

Source   Kotler, P., *The New Competition*, Prentice-Hall, Englewood Cliffs, NJ, 1985.

Example   **3.5(b)**

Subject   **Selling offshore equipment in Brazil**

Background   Brazil is the largest country in South America, covering 48 per cent of the
                 continent. Of its 8.5 million square kilometres of area, 36.5 per cent
                 consists of sedimentary basins and 9.4 per cent is continental shelf. One of
                 the country's major activities is the production of oil and gas.

                 Following the 1990 election, President Collor's administration opened
                 Brazil up to imports and initiated a programme of privatization. The state
                 oil company, Petrobras, the largest company in the country, was thus
                 allowed to buy goods and services from overseas. This in turn offered new
                 market opportunities to UK suppliers of oil and gas equipment who have
                 gained considerable experience in the offshore sector through servicing
                 operations on the UK continental shelf.

                 The Scottish Development Agency (SDA), in its role of assisting Scottish
                 industry, has produced a booklet giving the background to Brazilian oil
                 and gas activities. The section devoted to 'Doing business in Brazil' deals
                 with routes for market entry and indicates nine points of importance, as
                 follows:

- *Market contact*:   Personal contact is regarded as essential in order to
  achieve commercial involvement, because recognition in Brazil cannot
  be presumed on the basis of a reputation with North Sea operators. The
  SDA recommended that interested Scottish companies should take part in
  the oil show held in Brazil, because this attracts all 32 South American oil
  companies.
- *The client*:   Petrobras would be the client and since it is a very large
  organization, a multi-level approach would be required, which should be
  preceded by careful market research.
- *Possible niches*:   Most oilfield equipment used in Brazil is produced
  locally, often under licence from US companies. There are, however,
  opportunities for more specialized products which are competitively
  priced.
- *Points to be emphasized*:   The key factors to be kept firmly to the fore in

all business dealings are product quality, cost benefits, and technology that is new to Brazil.

- *Agents/distributors*: Overseas suppliers need to have a local agent who can handle the importing of their products, provide useful market intelligence and generally smooth the importer's path.
- *Business mechanisms*: A number of approaches are possible, such as selling or manufacture under licence. Advice should be sought from local experts before selecting the particular approach for each case.
- *Joint ventures*: A joint venture is one very useful method of working, and greater mutual understanding can be built up between internal and external organizations as they learn to work together. General information regarding suitable partners and the experience of other companies can be obtained from the British Consul General in Rio de Janeiro.
- *Finance*: In many cases it is necessary to have up-front financing when bidding for big equipment contracts, and it is essential to arrange this well in advance.
- *Approach*: Companies were advised of the need for patience combined with persistence in order to be successful.

Comments   For companies unfamiliar with countries in South America, such as Brazil, it is useful to have some background data to serve as a general guide. In this case, a political change has led to fresh opportunities, but even the best-known suppliers to the North Sea offshore market will have to work hard just to gain a foothold for their products in a potentially very large market.

Source   *Brazilian upstream oil and gas market*, published by the Scottish Development Agency, Aberdeen, October 1990.

Example   **3.5(c)**

Subject   **Image and logo**

Background   A first step towards effective selling is to establish the correct image for the company and to achieve customer identification. In other words, it is to make sure that potential customers can readily recognize the company and its activities. In this connection the choice of a logo and other symbols is extremely important. Many companies have devoted a significant amount of managerial time and resources to ensuring that the correct 'corporate image' is promoted.

Ideally, the company's logo or symbol should have the following key features:

- Instant recognition

- Easy to remember
- Attractive
- Projects a dynamic image

A sample set of logos is given in Fig. 3.3. Readers are invited to judge for

British
TELECOM

PHILIPS

Figure 3.3 Some familiar company logos

themselves how well these criteria have been met by the company in each case.

- *BMW*:   A German manufacturer of high performance cars, with its headquarters in Munich. The symbol represents a range of prestige vehicles.
- *Philips*:   A European company which is a leader in electrical products and developed the compact disc (CD) system for recording music.
- *British Petroleum*:   A multinational oil company, often referred to simply as BP. In 1988, at significant cost it updated its familiar green shield for a new logo in order to project a more dynamic image.
- *Hewlett Packard*:   The company describes itself as follows: 'An international manufacturer of measurement and computation products and systems recognized for excellence in quality and support'.
- *Rolls-Royce*:   An engineering company which builds high performance engines for many types of aircraft and power stations. Its name originated with the famous luxury motor-car of the same name.
- *BT*:   This is the new trading name of British Telecom. The symbol is described by the company as representing 'two human figures, one listening, one speaking, brought together by BT's technology and understanding of customers' needs'. The cost of producing this new identity was quoted in the press at over £60 million!

Comments   The customer's perception of a particular logo or symbol may not be readily defined and to a large extent it is subjective. There is, however, no doubt that a logo which meets the criteria indicated above will greatly increase customers' affinity with the company's products. Expert assistance is necessary in order to develop a good logo, and this can prove very costly.

Source   Manufacturers' brochures. Permission to use the logos is gratefully acknowledged.

3.6 INVENTION AND INNOVATION

Goal

To enhance the performance of an organization through potentially profitable improvements to existing—or the introduction of completely new—products, systems, procedures or services.

About the subject

Although there is a clear distinction between their meanings the terms 'innovation' and 'invention' are often treated as synonymous, because the division is not always clear-cut in practical situations. 'Innovation' is the adaptation or extension of some *existing* technology, method or procedure in order to improve the performance of a system, process or production function. 'Invention', on the other hand, is the conception and implementation of a completely new response to a practical problem or perceived need, leading eventually to a new product.

Every form of technology is initially an invention which eventually becomes routine technology, but in the course of time innovation is called for to improve or extend its performance in the light of new demands.

From a business point of view it is essential to decide how much of the available resources should be devoted to invention and how much to innovation. Since it uses existing technology, innovation involves less risk and stands a better chance of success. It is also likely to give a speedier return on investment. Invention, on the other hand, usually takes longer and costs more, although in the long run it may provide a better return. Equally, it may lead to nothing because of unforeseen problems in making the new product commercially viable.

Key questions and issues

How does invention come about?

An inventive idea can be generated by anyone, for a variety of reasons. Ideas that lead to inventive products stem from the following main sources:

- Companies already active in an industrial area devoting research effort directly towards coming up with new products
- The result of studies by research organizations an higher educational institutions

- Individuals who are knowledgeable about a particular subject and have experience of coming up with new ideas

Implementation of an invention

There are a number of ways of taking a good idea through to the stage of a product that will sell successfully on the market. Twelve basic steps are generally needed to convert an inventive idea into a viable product for commercial exploitation, but the second six in invention are identical with the six in innovation. The first six steps in invention are summarized in Table 3.4. They are as follows:

- *Identify*: Some form of market research study should be done to identify areas requiring invention so that their commercial potential can be established.
- *Generate*: The next step is to generate original ideas in response to an identified need. After the first oil crisis in 1973, for example, many inventive ideas were put forward for exploiting alternative sources of energy.
- *Verify*: A lack of awareness of previous work can result in 're-invention'. Inventive ideas, therefore, have to be verified to ensure that they are indeed original. This can be done by means of a search on similar products and by checking existing patents in the subject area.
- *Select*: It is necessary at this stage to select the most promising idea from those generated. Too many possibilities inhibit concentration on a specific way forward.
- *Check*: A physical model should be constructed to demonstrate the idea and check its practical feasibility.
- *Patent*: Once it is ascertained that the novel idea is an original one the features associated with the product, process or device should be protected by means of a patent.

It is clear that the road to success in invention is long and full of risks.

Table 3.4   A route towards invention

STEP	ACTION	TASK
1	Identify	problems offering inventive opportunities
2	Generate	original ideas
3	Verify	their originality
4	Select	the most promising idea
5	Check	its feasibility with a physical model
6	Patent	the product for protection purposes

Special features of invention

In seeking to implement an inventive idea, the following factors should be particularly noted:

- *Low success rate*:   Inventors in every field generate huge numbers of original ideas but only about 2 per cent will achieve successful commercial exploitation.
- *Extended research period*:   Converting a viable idea into a commercial proposition usually involves considerable research and development effort. It may take several years before the new invention can reach the market.
- *The question of patenting*:   New ideas and the resulting products, devices and processes can be plagiarized by rivals of the initiating individual or organization with a consequent loss of benefit or profit to the latter. This is prevented, at least for a time, by means of a patent. Patenting is an expensive activity and is usually done via a lawyer or patent agent. It should be noted that acceptance of an idea's originality can be prejudiced if information on it is published before a patent is awarded.
- *The timing of the patent*:   Once the details of a patent are published, fear of legal action for an infringement of patent should prevent rival concerns from imitating or modifying the idea. It is to be noted, however, that legal action may not be successful. Patenting needs to be carefully timed in relation to the proposed date for starting to sell the invention.
- *Managerial suspicion*:   The management of a business may not necessarily understand all the advantages of a new and original idea. It will often weigh the estimated cost of implementation against the uncertainty of commercial success. An inventor may well have to overcome considerable resistance before an idea is implemented. It is important for an inventor to be good at selling as well.

How is an innovation initiated?

Innovative action usually stems from a need to improve the quality of a company's product while keeping the cost of manufacture at or below the current level. If successful, it will give the company a competitive edge over rivals, with the least financial and technological risk. The emphasis, however, must be on a quick return for the necessary investment, with the minimum of technological alteration or addition. The following are typical benefits from successful innovation:

- More effective use of resources in a production process
- Improved quality control
- The transfer of a successful technical solution to another application

Innovation can also serve as the second stage in the development of an invention.

The basic steps in innovation

The approach to a particular innovation will depend on the product, system or device concerned, but introducing any innovation involves the following six basic steps, which are summarized in Table 3.5:

- *Evaluate*: Perform market research to identify possible improvements desired, and evaluate the special features of rival products.
- *Quantify*: Quantify the aspects of the product, system or device that could be improved. It is at this stage that the scope of work involved should be defined.
- *Seek*: Seek financial support for the proposed innovation within the organization and/or from outside sponsors.
- *Solve*: Overcome problems by doing development or research work.
- *Construct*: Build the modified and improved version of the product for test purposes. Further modifications may be necessary in the light of the experience gained here.
- *Prepare*: Prepare for the introduction of the product or service to the market.

Table 3.5   A route towards innovation

STEP	ACTION	TASK
1	Evaluate	market opportunities for inventions and improvements
2	Quantify	work scope
3	Seek	funds and sponsors
4	Solve	outstanding problems by doing research or development work
5	Construct	a prototype for testing
6	Prepare	for introduction to the market

Special features of innovation

The significant aspects of innovation to be noted are as follows:

- *Small changes*: Since the changes required will usually be relatively small and based on existing technologies there is a good likelihood of success.
- *Speedier development*: Innovations normally require more development than research attention and can usually be implemented much more quickly than inventions. A return on capital invested is also likely to be achieved within a shorter period.

- *Managerial support*: Because innovation uses known technologies and methods its results are more readily understood by the senior decision makers in an organization. Hence, if an adequate marketing case is made, the necessary funds for the development work will probably be forthcoming.

Which should we encourage?

An organization may have to choose between a policy of promoting innovations and backing a specific invention. As indicated above, the tendency in general is to support innovation, and the case for an invention can be hampered by the fact that few inventors are commercially orientated and many are poor communicators. However, their inventive proposals should not be rejected simply because of their personal limitations as sellers. In order to arrive at a sound decision on such matters the following factors should be taken into consideration:

- *Objectives*: The basic task must be to determine whether or not the invention fits in with the organization's objectives and strategy.
- *Initial study*: Once it has been decided to examine an invention more closely, resources must be committed to that end. For example, time must be made available for studying and making a critical assessment of the proposal, and finance must be provided for the construction of a physical model, patent application etc.
- *Decision deadline*: A realistic date should be set for deciding whether to continue with or terminate the project. By that time the commercial and technological hurdles to be overcome should have been identified and quantified.

ILLUSTRATIVE EXAMPLES

Example	**3.6(a)**
Subject	**Examples of invention**
Background	Certain inventions have proved to be particularly successful in commercial terms and a brief examination is made here of some which have made their mark during the second half of the twentieth century:

- *Hovering systems*: The hovering concept was first applied to the design of a vehicle which is a hybrid of a ship and an aircraft, i.e. the hovercraft. The principle found its greatest commercial success, however, when it was applied to other activities. Successes include the lifting of heavy objects to transport them to new locations over a solid surface. Ease of operation has also been increased in cases such as the hovering lawnmower.
- *Optical fibres*: Optical fibres are thin, light, flexible and resistant to high temperatures. They are being used increasingly for the transmission of large quantities of information at very high speeds using light beams. This

has completely altered the way in which communication systems are designed.

- *Infrared devices*:   Heat-sensitive devices have been invented which can produce images of objects even in darkness. These are ideally suited for detection in situations where other methods have proved unsuccessful. Possible applications include the detection of aircraft and burglar alarms.
- *The compact disc*:   Greatly improved sound quality is possible through storing music on optical discs and playing it back through the use of a laser. This system does not suffer from the problems of distortion encountered in conventional disc and needle or magnetic tape-recorder systems.

Comment The features common to all these inventions are the simplicity of the basic concept and the existence of large markets for the applications developed. Once a mass production process has been perfected for inventions of this type, commercial success can usually be expected.

Source Nayah, P.R. and Keggeringham, J.M., *Break-through*, Mercury Books, London, 1987.
Press comments.

Example **3.6(b)**

Subject **Examples of innovation**

Background Many innovations have achieved commercial success and one or two that have made their mark in the second half of the twentieth century will be discussed briefly.

- *Containerization of goods*:   For many years goods transported by land or sea were treated 'individually' from starting point to final destination. Efforts to improve efficiency in transportation led to the concept of an integrated system using a standard size of container on road and at sea. This in turn led to the development of specialized lorries and container ships. The result was considerably increased efficiency in the transport of goods and a notable reduction in costs.
- *Recycling of waste heat*:   Heat is required in many manufacturing processes but much of it is often lost into the air. It is possible to reclaim this waste heat by designing the entire system so that it is recycled, thus making more efficient use of each unit of energy. When such recycling is adopted the gains are both environmental and commercial. Pollution is reduced, and the associated reduction in energy costs will also lead to a cut in the cost of the final product.
- *Introduction of robots in industry*:   Many manufacturing operations are repetitive and involve relatively simple tasks. By integrating machinery

with sensors and computers, robotic devices can be designed for such tasks that will give accurate and reliable results, particularly in the assembly line. This can lead to higher quality production and possibly an overall reduction in the unit cost of products.

Comments   The common feature of these innovations is the impact they have had on commercial activity without the use of new technology. The development time was much less and the chances of success were good in each case.

Source   Press comments.

Example   **3.6(c)**

Subject   **Patenting an invention**

Background   An important aspect of the process that takes an inventive idea from conception to actual product is the application for a patent. Many people with inventive ideas find this a complicated process, and the UK Patent Office has recently produced an explanatory booklet on its role and activities for the benefit of intending patentees. This contains much useful information for engineers involved with developing an invention.

According to the UK Patent Act 1977, the following conditions have to be satisfied before an invention can be patented in the UK:

- *The invention must be new*:   This implies that the invention has not yet been made in the UK (although it may have been patented in another country already).
- *There should be an inventive step*:   In other words, the invention should be more than an extension of an existing approach that would be obvious to someone with good knowledge and experience in the subject area.
- *There should be a practical indication*:   The inventive idea must be shown to be capable of being made into a product, process or device.

It will be evident from the above list that a number of inventive acts are excluded from patenting, and these include an idea, a concept, a mental process and a computer program. The last-named may, however, be protected by registering the developer's copyright.

There are three basic features to be noted once a patent has been granted:

- The patent-holder has a monopoly to exploit the invention for a given period, i.e. to make, use or sell the invention for a certain number of years. In the UK this period is 20 years from the date when the patent was first filed.

- Fees have to be paid for the initial filing of a patent, and for its subsequent maintenance.
- Details of the invention will be disclosed by the Patent Office through its publications.

Publication makes the technological information contained in the patent available to other inventors and researchers, who will benefit in the following ways:

- Gaining an insight into ways in which the problems they are facing have been tackled by others
- Being able to check that ideas on which they are working have not been patented already
- Finding inspiration for possible future inventions
- Identifying what competitors are doing
- Obtaining information of personal interest

Relevant literature, such as the Patent Office booklet, will assist inventors towards the successful filing of a patent and avoidance of possible pitfalls on the way. Since the process is both lengthy and costly, however, it should be remembered that in certain circumstances it is better to gain commercial benefit from use of the invention prior to the granting of the patent. This action does involve the risk of plagiarism but can have certain commercial advantages.

It should also be noted that many inventions which failed to obtain a patent have been treated as innovations and achieved significant commercial success.

Comments   The procedure for patenting requires special expertise which most inventors do not possess. However, a basic understanding of what is involved will ensure that the inventor of a new process or device can benefit fully from this protection.

Source   *The Patent Office*, UK Patent Office publication, 1983.

3.7 MATERIAL FOR FURTHER STUDY

Baker, M.J., *Marketing*, Macmillan Education, Basingstoke, 1982.
de Bono, E., *Opportunities*, Penguin, London, 1986.
Borg, J., *Selling yourself*, Heinemann, London, 1989.
Davidson, H., *Offensive marketing*, Penguin Business, London, 1987.
Drucher, P.F., *Innovation and entrepreneurship*, Heinemann, London, 1985.

Kotler, P. and Armstrong, G., *Marketing: an introduction*, 2nd edn, Prentice-Hall, Englewood Cliffs, NJ, 1990.

Lawrence, AS., *Fundamentals of industrial quality control*, Addison-Wesley, Wokingham, 1986.

Luck, D.J. and Ferrell, D.C., *Marketing strategy and plans*, Prentice-Hall, Englewood Cliffs, NJ, 1983.

McDonald, M.H.B. and Leppard, J.W., *Effective industrial selling*, Heinemann, London, 1988.

Oakland, J., *Total quality management*, Heinemann, London, 1989.

O'Hare, M., *Innovate*, Basil Blackwell, Oxford, 1988.

Roberts, E.B. (ed.), *Generating technological innovation*, Oxford University Press, Oxford, 1987.

Simmonds, K., *Short marketing case*, Philip Allan, London, 1987.

Smith, N.I. and Ainsworth, M., *Managing for innovation*, Mercury Books, London and W.H. Allen, London, 1988.

Twiss, B., *Managing technological innovation*, Pitman, London, 1986 .

West, A., *Course in marketing techniques*, Paul Chapman, London, 1988.

CHAPTER 4

MANAGEMENT

Certain aspects of business activity are primarily managerial responsibilities, and six of these are considered here.

The first section is devoted to an examination of *objectives and strategies*, since the selection of the right objectives and the devising of appropriate strategies for achieving them is vitally important to the success of any business, and indeed any other activity.

The second section is concerned with *contracts* and what they involve. Areas of potential dispute and the need for precise specifications are highlighted. Not only contracts but all business activities involve *negotiation* and so the importance of acquiring skill in negotiation is considered next. The basic elements of negotiation and the factors influencing its effectiveness are highlighted. The basis of *project management* is introduced in the fourth section, which summarizes the key techniques necessary for its implementation, and discusses the role of the Project Manager.

Since modern businesses depend very much on a flow of *information* and its effective management, a section is devoted to this topic. Points considered include suitable gathering and storage procedures to be employed, the contribution of the computer and problems associated with the maintenance of and right of access to the information stored in the database.

The *research and development* (R&D) policy of a company is the responsibility of the management, and this topic has therefore been chosen as the sixth in the set. After 'development' had been distinguished from 'research', attention is given to such features as the amount of R&D to undertake, and the generation and prioritization of R&D ideas.

It should be noted that the six topics selected do not cover every aspect of management. In particular the specialized area known as operations management has not been dealt with. The reader is advised to consult published texts for information on this topic.

4.1 OBJECTIVES AND STRATEGIES

Goal

To provide overall direction for an individual, an organization or an activity, and indicate the necessary series of steps to reach the target selected.

About the subject

The main purpose of an objective is to define the basic direction to be taken by an organization, team or individual. It may consist of a specific target to be reached, the successful taking up of an identified opportunity, or a means of focusing attention on an issue. Alternative terms used include: goal, aim, target and destination. A clearly defined objective will help to promote higher efficiency, improve the utilization of resources, encourage effective teamwork and assist in the meeting of deadlines.

It is a real challenge to devise an objective which is both ambitious and achievable. To formulate a balanced objective requires knowledge, understanding, experience and vision, together with a positive attitude on the part of those involved. Once it has been defined, agreed and understood, a series of logical steps must be drawn up and properly implemented to ensure its achievement. Each stage should incorporate certain targets to be reached with specific periods, and the approaches or policies to be adopted are also sometimes indicated. The finalized series of steps, when properly linked, is usually termed the 'strategy'.

Objectives and strategy are considered together because they are interdependent concepts. The efficient achievement of an objective is unlikely without a plan of campaign, but a strategy would be meaningless unless directed towards a specified objective.

When an organization links together an objective and its associated strategy the term often used is 'corporate strategy'. In effect this is a plan for a medium-term or longer period showing intermediate targets and the steps necessary for achieving them. Private individuals often draw up similar sets of objectives and strategies for periods ranging from one to five years.

Key questions and issues

Why is this subject important?

An objective provides a direction, and a strategy provides a method of achieving the objective. Both are vital for several key reasons.

Firstly, organization and individual alike must make efficient use of all the available resources in order to perform their various functions. Typical examples of these would include: effective use of time by the staff and investment in up-to-date facilities. None of this can be done in the longer term without clear direction and logical methods.

Secondly, objectives and strategies provide motivation for everyone concerned and a focus for the development of loyalty.

Thirdly, clear objectives will enable everyone to make positive contributions. Examples of the value of such objectives include the effective meeting of deadlines and their use as a yardstick for measuring the performance of individuals and groups within an organization.

Fourthly, properly defined objectives and supporting strategy will project the desired image, thus allowing people to see whether the organization is one with which they would happily do business.

Devising objectives

Devising an objective for an organization may seem a fairly straightforward task, but in practice it can be one of the most challenging of all. And the devising of a *balanced* objective is a real difficulty. It is counter-productive to set goals that are impossible to achieve and equally unhelpful to set them too low. The selection of an objective for a particular activity usually depends on market conditions, but it also involves the vision, knowledge and experience of key personnel. Care also has to be taken to avoid confusing the objective itself with the methods for its achievement. For example, if the *objective* is to fit a sensor on to an engine to indicate when a specified temperature level is reached, the crucial factor is the specification of the sensor's performance. Details of the selection process and the installation of the sensor are part of the *strategy*, and factors such as size, shape, weight, range and location within the engine are all subsidiary aspects of performance. It will only confuse the issue if these details are included in the definition of the objective.

A procedure for defining an objective

A possible procedure for defining an objective involves six basic steps is illustrated in Table 4.1. There are a number of key factors to note. First of all, a critical review of information obtained through market research is likely to yield a vast number of options, some feasible and others quite unsuitable. Secondly, a portfolio of suitable objectives should be generated and each examined carefully in order to select the most suitable one. Thirdly, the final choice should be based on a combination of analysis and experience gained in previous evaluation exercises.

Table 4.1   A procedure for defining an objective

STEP	ACTION	TASK
1	Review	market research findings
2	Generate	a portfolio of possible goals
3	Identify	critical factors and decision criteria
4	Evaluate	the suitability of all possible goals
5	Check	whether there are alternatives
6	Focus	on the most suitable goal

The main elements of a strategy

In devising strategies it is helpful to refer to the past experience available in the organization, in particular to ensure that those devised are neither based solely on unlimited enthusiasm nor too greatly inhibited by an over-cautious outlook. A given strategy may be short-, medium- or long-term in nature but should always be closely linked to the objectives concerned. Its basic elements will include at least some of the following:

- *Assessment*: A critical assessment must be done to determine the strengths and weaknesses of the organization or individual concerned. This should be considered together with their ability to take up available opportunities and deal with threats from competitors or other constraints. Methods used include the SWOT technique (Strengths, Weaknesses, Opportunities, Threats), and the Boston Box.
- *Planning*: Planning is needed to ensure that the sequence of actions is logical and that appropriate intermediate targets have been established. The practical implications of the strategy should also be examined carefully to quantify potential difficulties.

● *Testing*: Computer-based simulation techniques can be used for testing and assessing the steps in a proposed strategy, and to provide feedback to those involved in planning.

A procedure for devising strategies

Since the strategy is dependent on the objective selected it is very difficult to generalize here, but an outline procedure involving six basic steps is provided to give some broad guidance (see Table 4.2). Three features call for attention:

● It is essential to understand fully the selected objective for which a strategy is required.
● Background information and capabilities and constraints should be carefully assessed before a plan is derived.
● There should be facilities for regular reviews, and for refinements to be incorporated if necessary.

Table 4.2   A procedure for deriving a strategy

STEP	ACTION	TASK
1	Understand	fully the objective
2	Audit	critically the background information
3	Assess	capabilities and constraints
4	Prepare	a plan with defined stages
5	Verify	the plan's overall feasibility
6	Document	the procedure involved

ILLUSTRATIVE EXAMPLES

Example	**4.1(a)**
Subject	**Objective and strategy for an engineering company**
Background	This example highlights the identification of an objective and the derivation of a strategy in the early 1980s for the research unit of a company with general engineering and marine interests.

Identification of objective
The steps involved in developing the objective were as follows:

● *Review of market research*:   The data obtained in a recent market survey were considered along with areas of research potential. It was agreed that, in view of the potential offered by North Sea oil production, the unit's attention should be directed at oil-related activities.

- *Portfolio of possible goals*:   The three main possible options identified were as follows:
  (a) To solve short-term problems for the oil industry on a consultancy basis
  (b) To undertake strategic research projects of less than twelve months duration on problems of relevance to engineering
  (c) To concentrate on more fundamental and longer-term tasks on a joint funding basis in cooperation with oil companies and government departments

- *Critical factors and decision criteria*:   A number of critical factors were identified in relation to any potential project, the key ones being: the availability of contacts; bidding methods; time-scales; staff capabilites; constraints; project management skills; the facilities required; competition; methods of identifying problems; selling skills; problems associated with joint funding; contract conditions. A set of criteria was derived to assist in decision making.

- *Evaluating the goals in the portfolio*:   Each goal in the portfolio was examined in turn with the aid of the derived decision-making criteria, and lists of advantages and disadvantages were compiled for each option. The conclusions drawn in each case can be summarized as follows:
  (a) Short-term problem-solving was very attractive from the point of view of the company's understanding of the industry, the selling of its services and possible earnings. However, there were already many consultants working in this sector.
  (b) Strategic research was also very attractive for the same reasons as the short-term problem-solving option. In addition, the longer time-scale available in this case would be appreciated by the staff doing the work. However, the competition here was intense and it takes time to create good working partnerships.
  (c) The long-term research option was the one with most appeal to the researchers and engineers and they felt it was a good way of making useful contributions to the industry. Its principal drawbacks were the reluctance of the industry to invest in long-term research and the limited willingness of government sources to support any project which had low industrial funding.

- *Consider possible alternatives*:   Examination of the three options had not resulted in a firm decision in favour of any single one, and so alternatives based on a combination of two or even three were explored.

- *Focusing on the preferred goal*:   It was decided that the unit had to be involved in the total spectrum of activities ranging from short-term problem-solving to long-term research and the finalized objective was stated as follows: 'To develop a balanced portfolio of projects with industrial and national relevance, comprising fundamental investigation, medium-term studies and problem-solving assignments'.

Strategy

Once the above process had been completed a strategy was developed by means of the following steps:

- *Understanding the objective*: The first step was to ensure that the objective was clearly identified and understood in its entirety.
- *Auditing the background data*: A critical review of all relevant background information was carried out with particular reference to the following: the industrial sector in which the business has to operate; the locations of markets; leading organizations in the sector; the identities of competitors and their policies; short- and longer-term economic forecasts; future technical developments.
- *Assessment*: A critical assessment was done to provide an audit of the company's strengths and weaknesses in relation to markets, resources and management, and also to identify opportunities and threats.
- *Preparing a plan*: A plan was then drafted for the next five years, with targets for each three-month period. This involved both the overall performance of the unit and specific aspects such as finance.
- *Verifying the plan's feasibility*: The plan's overall feasibility and, in particular, its technical and financial projections were checked by a number of methods, e.g. independent assessment by a consultant and computer simulation. Modifications were made in the light of these findings, e.g. a re-assessment of the time needed to complete some of the projects in the initial period. Facilities for regular reviews were also incorporated into it.
- *Documentation*: The documentation for the 'best' strategy at that point was prepared and circulated to the relevant personnel. Its key elements included:
  (a) Market research studies on a six-monthly basis
  (b) Allocation of resources for problem-solving and medium- and longer-term research
  (c) The building up of three teams of core staff within the first year, with each team responsible for the selection of its own project work and its financial management
  (d) The introduction of regular training sessions as part of the staff development programme. Topics covered would include selling techniques and communication skills.

Comments   It is not possible to provide more detail on this exercise, except to note that formulating the objective took twice as much effort as deriving an appropriate strategy did.

Sources   Private communications.

Example	**4.1(b)**
Subject	**Objectives and strategy of a major UK company**
Background	Prior to an Extraordinary Meeting of a major UK engineering company to gain shareholders' approval for a takeover action, a document was circulated which summarized the Directors' objectives and strategy for the 1990s. These are highlighted here.

Objectives

This section began by stating that the directors had recognized for some time that they must take further steps to meet the objectives of the company and that the proposal before the meeting stems from that fact. It then reiterates their principal objectives, which are: 'Achieving leading positions in world markets for our products, strengthening and growing our core businesses and providing a substantial and sustained increase in earnings for our shareholders.'

Main elements of strategy

There were seven points here, of which the most important are:

- 'To develop a substantial and profitable presence in European markets in advance of 1992 through acquisitions and international alliances in our core businesses'
- 'To expand our operations in the United States and elsewhere overseas'
- 'To satisfy our customers by offering the most technically and cost-effective solutions to their needs'
- 'To ensure technological leadership in our core businesses'
- 'To invest our surplus cash resources profitably in business which will complement and contribute to our on-going activities'

This statement was followed by sample results of the application of this strategy and the consequent improvements in earnings and dividends.

Comments	Information of the type given in this document is not always readily available. When one company is bidding to take over another, however, it can be expected to issue such statements for the purpose of impressing the shareholders in the targeted company. It would normally be difficult to find fault with the content of these documents. However, it would be interesting to follow the development of the amalgamated companies over a period of three to five years to see whether the stated objectives have been achieved and to assess the effectiveness of the proposed strategy.
Source	Company brochure of the General Electric Company, UK, when taking over Plessey.

Example	**4.1(c)**
Subject	**A lesson on objective and strategy decisions**
Background	During the period 1970 to 1990, the major UK oil companies all selected new objectives and revised their strategies in order to maintain a high level of performance. Some of these new objectives led to successful business activities, while others resulted in expensive failures.

Sir Peter Walters, who was the Chairman of British Petroleum (BP) between 1981 and 1990, highlighted the selection of diversification as the goal for BP and the lessons learnt from the experience.

Most oil company leaders were influenced by the first oil crisis of 1973 when many of the oil producing countries nationalized the foreign companies which were producing their oil and a number of the 'home' governments were seeking to regulate the activities of an industry which has a key role in any nation's economy. BP, like a number of other major oil companies, decided that diversification was the objective to adopt and in the mid-1970s it acquired companies in the coal, minerals and information technology industries. The case for their decision was the belief that these activities were a 'natural' extension of the oil business and that BP could benefit from the resulting synergy. Once involved in these industries, however, BP made some disconcerting discoveries, as follows:

- *Coal*: They did not have an adequate resource base or market share, and there was no real opportunity to develop a special identity for its product.
- *Minerals*: The company did not have the right expertise and experience for this type of exploration.
- *Information technology*: They were experienced in the use but not in the supply of computers.

By the early 1980s, BP was experiencing losses in chemicals, shipping and the European oil business.

This situation led Sir Peter Walters to examine every business within the BP Group, by applying two searching criteria.

- *Critical mass*: The question to be asked here was, 'Could this business achieve a big enough share of the market to compete effectively with others in the field?'
- *Selective excellence*: The question here was whether the quality of the product/service was acceptable.

If the business could satisfy both these criteria, its objective was to become one of the top three companies in the field over a given period. In the light of this decision BP had two options for every activity: either double the size of the business or leave. By adopting this strategy the company was able to rationalize its activities during the 1980s by getting out of the coal industry and selling its mineral interests and information technology business. BP then prospered by concentrating on its oil business.

Comments   This case clearly demonstrates the importance of having the correct objective and adopting the right strategy. It should be remembered, however, that companies tend to follow trends, and diversification was popular in the 1970s. Many companies had this objective and found to their cost that it did not ensure profitability.

Source   Article by Sir Peter Walters in the *Independent on Sunday*, 9 December 1990.

4.2 DEALING WITH CONTRACTS

Goal

To define a working, and preferably legally enforceable, arrangement for all the parties involved in a business transaction or commitment.

About the subject

A contract is a formal agreement between two or more parties—individuals, organizations or even countries—which is usually enforceable by law. It is, however, important to define the legal system or systems under which it is to operate, as different countries have different legal provisions. A typical example of a business contract would be an agreement covering the supply of goods or services by one party to another at an agreed price. A detailed description of what is to be supplied will usually be included in the document.

Contract documents can vary considerably in form, depending on the type of task, duration and cost of the activity involved, and the relationships between the various parties. At their simplest they may be a letter or even a telex from one responsible person to another, but the contracts for engineering projects are usually extremely complex documents. Such complexity arises from the fact that engineering projects are likely to involve many stages and more than two parties. The amount of work involved in preparing a contract is directly proportional to the number of parties involved.

The following features must be included in order to provide an effective contract for an engineering project:

- The specifications of the product or service to be supplied
- The approach to be adopted for meeting the specifications
- The schedule and detailed work programme
- Financial arrangements such as costs, method of payment and how the work is to be financed
- Legal aspects of the agreement and their effects on the project
- Liabilities for failure or poor performance
- Responsibilities of the different parties involved

Careful attention needs to be paid to each of these items for the greatest benefit of all the parties involved.

Key questions and issues

Why make formal contracts?

There are two basic reasons why contracts are desirable. The first is that making a contract ensures that the objectives of the activity and the working relationships between the various parties are both defined. In most engineering projects the contract will detail the work to be done for the agreed reward, and specify a completion date. It will also define the responsibilities of all the parties involved. The second point is that a well drawn up contract can keep possible disputes to a minimum by a clear detailing of requirements and financial responsibilities.

These are the reasons why careful attention must be paid by all concerned to the wording of a contract. There can, however, be problems here when liabilities are involved or when the contract has to be translated into one or more other languages.

The role of specifications

In the case of an engineering project there will usually be a full set of specifications associated with the contract document. These give details of the work to be done and methods to be used and their preparation calls for specialist knowledge of the activities involved, together with experience and sound judgement. Every aspect of the work has to be defined as accurately as possible. Points to be considered could include, for example, the choice of manufacturing equipment, the qualifications of the welders and the type of paint required. The national or international standards to be met are also indicated here. It is thus hardly surprising to learn that the specifications for a major engineering contract could take several months to prepare and involve several volumes of documentation. However, a well-prepared set of specifications can save money for an organization, whereas specifications that are rushed out may lead to seriously reduced profits should a dispute arise.

The proposed approach

The approach proposed for adoption in order to fulfil the contract specifications must be carefully examined and evaluated. Solutions put forward, particularly those involving advanced technologies, must be checked for the following points:

- The technologies have a sound foundation and have been fully validated
- Potential difficulties and possible ways of overcoming them have been considered
- Alternative suppliers of key equipment have been identified
- The standards to be used are realistic and will also give the desired level of quality

Work programme

Once the specifications are agreed on, work plans and schedules can be drawn up, but, before they are finally settled, it is essential that all concerned should understand them and agree on their interpretation. Specialist knowledge, experience and sound judgement are thus also necessary at the review stage to enable the implementer to predict possible snags. Too rigid an interpretation of detail could prevent agreement being reached, but accepting loosely worded specifications could well result in disputes at a later stage.

Most contracts specify the period of time involved. A contract for a new product, for example, could include an actual delivery date. A contract for another type of assignment might require the work to be completed within twelve months of its commencement. In both cases a work programme is needed, indicating the various intermediate stages and the time needed to complete each of them. Such a programme facilitates the detailed planning of the project.

Financial arrangements

From the financial point of view three principal types of contract are used in engineering activities:

- *Lump sums*:   A fixed sum is agreed for a clearly defined scope of work.
- *Reimbursable costs with fixed fee*:   There are many variations under this heading but, in general, contractors are paid for their time only. In this situation the project specification is not precisely defined, and great care is needed to ensure effective control of costs and efficient work practices.
- *Measured*:   With this approach, use is made of a Bill of Quantities with details of the work scope. When pricing is based on a Bill of Quantities the cost remains fixed so long as the specification is not altered.

Methods of payment may range from handing over the whole agreed amount once the work is complete to the payment of a deposit on signing the contract followed by an interim payment on completion of each stage, or progress payments.

The financing of a project must also be considered, particularly a major project involving a large sum of money. The timing of payments and the method used can have a significant effect on the actual value of the money received and on profitability. The reason for this is that contracts are often financed with borrowed money and careful phasing can minimize the interest charges incurred.

Legal aspects

A contract is a legal document and its implementation will be greatly facilitated by careful preparation. It is essential to ensure that its provisions are understood by all parties, despite the necessary use of legal terminology. An engineer, for example, does not usually draw up a contract document but relies on colleagues with legal training. The engineer, however, must be fully aware of all its implications.

In addition, those responsible for the details of the contract should be encouraged to adopt 'sensible positions'. It is counter-productive, for example, for representatives without relevant technical knowledge to spend hours discussing terminology, as this could result in a delayed start to the actual work and consequent financial penalties for late completion.

Finally, where the parties concerned operate in several different countries, the contract must meet the legal requirements of every country involved so that necessary local action can be taken if a dispute should occur.

The use of penalty clauses

Most contracts include 'penalty clauses'. Their main purpose is to cover the question of liabilities for performance that does not meet the specifications and contributes to failure of some kind. These are not intended as punishments but provide a basis for the recovery of agreed costs through legal channels.

Typical questions covered by these clauses include:

- *Late delivery*: The supplier has to pay back to the customer a specified sum of money
- *Performance failure*: Agreed damages will be paid once liability has been established
- *Delays in payment*: An additional amount would have to be paid

Responsibilities

Conditions relating to the responsibilities of the different parties must also be defined. Depending on the type of project these may include:

- Ownership of results
- Exploitation and dissemination of results
- The procedure for reporting progress
- The agreed method of performing an audit
- Any special conditions

The aspect likely to involve considerable negotiation include:

- The quality of the finished product or service
- Details of the time-scale
- The price of the contract
- Liability if deadlines are not met
- Ownership of the results if new products are developed

ILLUSTRATIVE EXAMPLES

Example   **4.2(a)**

Subject   **Contracts**

Background   Two examples are used here to illustrate the importance of contracts and how contract discussion can get out of control.

The importance of a contract: On the basis of specifications provided by the railway company's engineer, a civil engineering company submitted a fixed price proposal for building a section of railway line. It was accepted, and the contract for labour, material and other costs was then signed.

When the work started it became clear that the railway company's engineer had underestimated the amount of earthwork by nearly two million cubic metres, and he tried, but failed, to make changes in order to reduce the additional cost. When work was finished the engineer certified for payment of the amount agreed in the original contract without making any allowance for the extra earthwork.

This led to a dispute that went to the courts. The Court of Appeal upheld the agreement in the railway company's favour because the contractor had agreed to build the section of railway line for a lump sum and the actual quantity of earthwork did not form part of the contract. The contractor had therefore to complete the requirements of the contract without receiving any payment for the additional work incurred.

Taking too long to draft the contract: In October 1989 researchers in an academic institution completed the research phase of work in the design of an underwater measuring device. This work was jointly supported by the UK Government's Science and Engineering Research Council and a major oil company. Before moving on to the development phase, with further financial commitment, the oil company's technical staff were keen to establish whether the device would operate effectively outside the laboratory environment by using it to take measurements on two specimens stored onshore. In October academics and the oil company's engineers agreed that these tests would be carried out over a three-month period between February and April 1990 for £5000 and a report would be produced. The results of this study would form the basis of discussion on a new contract. The 'arrangement' was initiated with a letter of intent from the oil company to the institution, and it was agreed that the contract would take the form of a 'service order' linked to the work programme already outlined. Several months later the necessary short-term contract was still not signed. The main problems were:

- The commercial sections of both the oil company and the institution were unable to agree on the wording of two clauses in the contract.
- The institution's representatives did not wish to sign a document which had implications for the patent rights of the product.

A financial settlement was finally reached after seven months of numerous long-distance phone calls, and several formal and informal meetings.

Comments

Firstly, the case of the railway line emphasizes the fact that every contract must be carefully checked to ensure that the calculations on which it is based are correct and complete. In this instance, even although the mistake was genuine, one party suffered as a result of the contract's wording.

Secondly, the difficulties between the academic institution and the oil company provide a typical example of the failure of commercial personnel to understand the technological objective of a project. By making a major issue of a very small sum they pushed the cost of agreement in this contract up to about three times the value of the contract itself.

Sources

South, V.P. and Soo, J.W., *Building contract claims*, Collins, London, 1983. Private communication.

Example

4.2(b)

Subject

Contracts with Commission of the European Communities

Background

Funds are becoming increasingly available from the European Community (EC) to support research and technological development activities by companies, research organizations and academic institutions. The majority of such projects involve partners from at least two states in the community. Before the Commission can provide its contribution to a project, formal contracts have to be drawn up for each of the partners involved.

For the guidance of the parties concerned in such projects, the Commission has prepared a model contract which conforms to the legal requirements of all twelve states. This model contract is a six-page document containing twelve 'articles' (or clauses) and three annexes. The twelve articles cover the following points:

- Scope of contract
- Duration
- Financial contribution of the Commission
- Payments by the Commission
- Cost statements
- Reports and deliverables
- Ownership and exploitation of results
- Technical verification and audits
- Amendments, variations or additions
- Special conditions
- Applicable laws and entry into force of the contract

These are followed by space for the signatures of the participants' represenatitives.

The three annexes contain detailed information about:

- The work programme
- General conditions covering the project: these range from the implementation of the work and the ownership, exploitation and dissemination of results to allowable costs and auditing
- Special conditions, if applicable

If this pattern is followed closely a smooth and fairly speedy conclusion of the agreement should be achieved and work can begin on the project itself.

Comments   This model contract is fairly standard but care must be exercised to ensure clear understanding of the general conditions by all concerned. This is particularly relevant to the questions of expected deliverables and the ownership of the results.

Source   *Model contract*, Publication of the European Commission, Directorate General XII (Science, Research and Development), October 1988.

Example	**4.2(c)**
Subject	**The procedure for seeking a contract**
Background	The process of winning a contract has a number of basic phases, regardless of the type of activity in question. This example considers the case of a particular mechanical engineering company, which manufactures a single unit or a small number of fabrications up to 1000 tonnes.

Phase 1: Initial enquiry   The company receives around a thousand enquiries per year, either directly as a result of past work or in response to a concerted sales drive. Initial enquiries are usually very general, indicating only the client's key requirements, and the potential value of the contract can range between £50 000 and £100 million. The company has documented a standard procedure for responding to enquiries.

Phase 2: Submission of tender   After discussion of each enquiry, the company will decide whether or not to tender. Generally, tenders will be submitted for around 20 per cent of enquiries received.

The submission will consist of the following:

- The scope of what will be supplied through the contract
- Conditions to be satisfied, data to be supplied by the client and the commitments of all parties
- Total cost of the contract
- Damage clauses in the event that the conditions are not fulfilled
- The terms of payment: a choice from fixed price, reimbursable at a given rate, or fixed price plus controlled price adjustment

Phase 3: Presentation   If the tender receives a favourable initial response, the client will usually invite the company and other potential suppliers to make individual presentations. The main purpose of this is to give an opportunity for the provision of additional information and for responding to questions.

Phase 4: Client's questionnaire   If the company remains on the short list of potential suppliers, it will be sent a very comprehensive questionnaire covering both technical and commercial issues in relation to the proposed contract. In the case of large projects this will probably arrive about two months after the presentation, with a request for its return within four weeks.

Phase 5: Contract negotiation   A series of discussions then follows on each clause of the proposed contract, and if these are successfully concluded the contract will be awarded to the supplier.

Comments   The purpose of this five-phase approach is to ensure that every detail of the contract is fully considered, and it is important that commercial aspects of the project be considered together with its technological aspects.

Source   Private communication.

4.3 APPROACH TO NEGOTIATION

Goal

To obtain the best possible deal for one's own organization in a business transaction.

About the subject

Negotiation is the process of discussion between two or more interested parties regarding a business transaction, with the aim of achieving an agreement.

There are many reasons why negotiation is both desirable and necessary, and some of the key ones are:

- A situation is not well defined, but can be clarified through constructive discussion.
- The conditions offered are not acceptable, but could be modified by negotiation.
- The contributions of several parties need to be pooled in order to achieve the optimum arrangement.
- Parties involved in an exchange of, e.g. special facilities, need to get to know each other better.
- Acceptable working relationships need to be established between two or more parties.

A good negotiator is one who is normally able to obtain the best possible terms for his side. These may consist of a discount, an extra facility or service, some fringe benefits or greater flexibility of operation. This should be true for *all* the parties involved in a set of negotiations—each one should be able to leave the negotiating table with something gained.

The art of negotiation is usually developed through experience of trade bargaining between companies and governments, but it touches the life of everyone today. Like many other business activities it requires skills which—once acquired—develop with use, and involves a number of processes, such as 'arguing' and 'proposing'. The vital factor is to keep focused on the main objective.

Key questions and issues

Observing a negotiation

An engineer observing professional negotiators in session may well be forgiven for failing to detect a pattern in the process. However, an observer is sure to be aware of the following features:

- *Arguing*: A representative of each party in turn presents the case for his or her side, highlighting its strengths and any weaknesses in the other side. Discussion at this stage can become quite heated.
- *Proposing*: At an appropriate point one party puts forward what his or her side would like and/or asks the other side to define their acceptable conditions. This stage can be an iterative activity involving a number of rounds of proposals.
- *Consultation*: Discussion may be suspended for a time so that each party can discuss their position in private, possibly in consultation with advisers. More senior members of the organization may be involved here, particularly if proposals from the other side are of an unexpected nature.
- *Bargaining*: Each side does its best to obtain concessions from the other, but if the gap between them is too wide the discussions may break down. This may cause one party to withdraw.
- *Deciding*: If an acceptable compromise is reached the decision will be taken to draw up and sign an agreement that is binding on all parties.

It should be noted that the total negotiation procedure involves all of the above stages.

Preparation for negotiation

Success in any set of negotiations depends on background preparation coupled with previous experience. Preparation takes different forms depending on the objective, the type of activity, the level of detail required and the importance of the business in question. It should cover at least some of the following aspects:

- Background information on the other party, such as policies and attitudes, the personality and style of the representative(s) and previous experience of negotiating with this party.
- The position of one's own side in relation to the matter under discussion, including such aspects as the degree of flexibility permitted, the time available, and the amount of background support that can be expected.
- Working out various options and devising possible negotiation strategies, on the basis of the information available.

A negotiation procedure

In order to become skilled, it is important to understand the basic elements of negotiation. It is, however, wise to have in addition a procedure which will take the disucssion step-by-step from its beginning to a successful conclusion. Such a procedure is particularly useful for introducing beginners to negotiation techniques and for testing their effectiveness in different situations. One procedure has been adopted by Gavin Kennedy and his colleagues and used successfully in international management training seminars (see Section 4.7). Table 4.3 summarizes the key steps of a modified version of this procedure, consisting of eight steps. It will not be necessary to use all eight in every case but, conversely, it may be necessary in some to introduce additional steps in order to achieve progress.

Table 4.3   A procedure for negotiation

STEP	ACTION	TASK
1	Prepare	by studying the background information
2	Present	the case by each side
3	Signal	the positions taken
4	Purpose	specific propositions for consideration from all points of view
5	Package	a draft agreement
6	Bargain	for possible concessions
7	Close	discussion and bargaining
8	Agree	to the deal

Two of these steps should be particularly noted. Firstly, it may be difficult to detect the techniques being used to indicate to the other parties a willingness to alter position and do business. These signals can take a variety of forms, for example, the use of 'coded' words. Secondly, skilful and well-timed winding up of a discussion is important in order to protect achievements from the possibly harmful effects of going back over previous ground to settle a relatively minor detail.

Attributes of an effective negotiator

To be a successful negotiator, particularly on the international scene, one needs a variety of attributes. The following five are, however, the most important:

- An understanding of the basic elements of negotiation and of how to apply them in given situations.
- A broad knowledge of a range of topics including the technical, financial and legal.
- The ability to think through the consequences of a proposal from either party on the spot and, if necessary, to suggest more feasible alternatives.
- The ability to recognize the strengths and weaknesses of both one's own side and the other party, and to apply psychology as appropriate.
- The willingness to commit adequate time and effort for thorough preparation before going into a negotiating session.

It must be strongly emphasized that anyone who wishes to be effective in negotiation at the international level must usually have had practical experience on home ground first. It is also necessary to know something about the other countries concerned. Otherwise there is always the danger of interpreting an innocent signal as a rude gesture!

Some causes of failure

Failure in negotiation can be attributed to one or a combination of the following causes:

- *Lack of system*: There are those who choose to ignore the basic elements of negotiating technique, and others who lack any understanding of what is involved in a negotiation situation.
- *No clear objectives*: Those on one side at least are unclear about what is wanted, and what will, and what will not, be acceptable.
- *Unrealistic position*: One side or the other may take up an unrealistic position and the gap between the aims of each side is too great to bridge.
- *Complaints in place of proposals*: In many instances negotiation sessions simply turn into exchanges of complaint. Regardless of whether the grievances are justified or unjustified it is essential to realize that positive proposals will help towards reaching an acceptable agreement.
- *Emotional involvement*: Success in negotiation calls for patience and coolness. Emotional involvement can lead to a stalemate or breakdown, through causing one to make statements with implications that go far beyond the original intention.
- *Indecisive position*: Sometimes one party acts indecisively and continually changes position during the course of the discussion. Others may put forward additional alternatives after the discussion has closed. The effect of this last is either to return the situation to the first stage, or to cause the other party to refuse to do business.
- *Rigidity*: Either because of personal style, or instruction from the organization, one

party may refuse to make any concessions at all and this prevents progress being made.

Improving negotiating techniques

In combination with a grasp of procedure several other factors can enhance negotiating technique, and these include:

- *Talking less but listening more*:   In many instances the negotiations have failed because at least one party failed to listen carefully to the terms or to recognize the signals offered by the other.
- *Flexibility*:   Within the scope of the negotiating plan it is essential to demonstrate a degree of flexibility so that the negotiations can proceed.
- *The relative value of concession*:   Specific concessions will have different values to different parties. It is essential to see an offered concession from the point of view of the other party and thus be able to judge its actual importance.
- *Face saving*:   In many countries it is important that both or all parties are seen to have 'won' in negotiation. There is therefore a need to ensure that the results of the discussions are tactfully presented.

ILLUSTRATIVE EXAMPLES

Example   **4.3(a)**

Subject   **Negotiation between a factory manager and a consultant**

Background   A multinational food processing company wished to expand the floor area of its factory in Paisley, Scotland, from 200 m^2 to 6000 m^2. The chairman of a small structural engineering consultancy was invited to discuss the project with the local factory manager with a view to his tendering for the structural work. The following outlines the steps in their negotiation.

Step 1: Prepare   The two parties had worked together before on another building project, but the way the architect had organized the work had left everyone unhappy with the operation of the contract. The client was, however, impressed by the standard of the consultancy's work and was keen to give it another opportunity to work for his company. At the same time, his own objective was to have the job done at minimum cost. The consultant's objective was, naturally, to win the contract, but he was determined to accept the job only if he received a 'fair' deal from the client.

Step 2: Present   The manager's case was based on the need to keep costs down because of his limited budget and, since he was prepared to give the consultant an opportunity to bid, he was expecting a 'good' deal. The consultant argued that he was able and willing to do the job but, in view of his previous experience, he would only be willing to take on the job if the reward was adequate.

Step 3: Signal   After presentation of both cases, the manager suggested that the consultant should put forward a price for the job and the consultant agreed that this was the way forward.

Step 4: Propose   The consultant estimated that the job would cost £1.5 million and said that his fee would be 3.5 per cent of this, which was in accordance with the scale recommended by his professional institution.

Step 5: Package   This was not considered necessary.

Step 6: Bargain   The factory manager's response to the quote was that he did not believe any consultant still used the 'official' scale for determining his fees. He indicated that the percentage was too high, and following a breakdown of the composition of the fee, suggested that a rate of 3 per cent would be reasonable.

Step 7: Close   The deal was closed with both parties agreeing verbally to a fee of 3 per cent of the total cost of the job.

Step 8: Agree   A letter was immediately prepared by the consultancy and sent to the manager to confirm their agreement with the outcome of the negotiation.

Comments   This incident illustrates a typical approach to negotiation in situations where the parties are known to each other. Although it was a very straightforward transaction, it involved seven of the steps in the suggested procedure.

Source   Private communication.

Example   **4.3(b)**

Subject   **Telephone negotiations for advertising space**

Background   Negotiations for advertising space are usually conducted by telephone with the initiative being taken by either the advertising client or the publisher's representative. Typical relevant advertisements would include those for:

- Vacancies for engineers in a manufacturing organization
- Engineering degree programmes available at an educational institution
- The announcement of a technical conference

The situation considered here is the renewal in 1990 of an advertisement for a special technical course on practical structural design in the journal of a professional institution. The steps involved in the negotiation were as follows:

Step 1: Prepare The client is targeting customers for the course among recent graduates in Civil Engineering and related disciplines. His budget for the purpose is £1000 and he would like to maximize impact with three or four insertions of an advertisement which has already proved effective. The journal's advertising linkman prepares by amassing data on rates, dates of publication and the degree of flexibility available.

Step 2: Present Having been informed that the 1990 rate for a single insertion would be £360 plus value added tax (VAT) at 15 per cent at the time in question, making a total of £414, the client argued that his was too much to charge on two grounds:

- It represented a 20 per cent increase on the previous year's rates, which was unreasonable.
- As a customer of twelve years' standing he was surely entitled to some special consideration.

The linkman responded that he had no choice, as this was the rate set for 1990. However, he was willing to let the client talk to the advertising manager in order to confirm the policy. The advertising manager justified the directors' decision to increase the charge to £30 per column-centimetre by rises in operating costs. He also confirmed that, at that rate, the client's 12 column-centimetre insertion would now cost £360.

Step 3: Signal In the course of further discussion the advertising manager indicated that he would like to continue the journal's good relationship with all its long-standing customers. The client was also keen to continue using a journal which had attracted a good response in the past.

Step 4: Propose The advertising manager proposed that the first insertion would be charged at the current standard rate, but offered a 20 per cent discount on the two subsequent insertions, bringing the cost of these down to £331 each, and making the total cost £1070. The client, however, was still keen to stay within his budget. He therefore decided to cut the length of the advertisement to 10 column-centimetres which, at a rate of £30 per column-centimetre, would bring the total cost of the three insertions below £1000.

Step 5: Package A possible arrangement was packaged for the consecutive insertions over the next three weeks at a cost of £345, £276 and £276, or a total of £897 including VAT. This left £103 in the budget.

Step 6: Bargain The client then asked whether a further reduction could be made so that four insertions could be achieved by means of a larger discount on the third and fourth insertions. The advertising manager could not accept this proposal but said he would be willing to put in an extra insertion at a later date if a vacant slot occurred.

Step 7: Close   The discussion was closed and the results of the discussion were then confirmed in writing.

Step 8: Agree   The deal was agreed.

Comments   This is an example of negotiation which satisfies both parties. The client maximizes the value of his advertising budget and sharpens the impact of the actual advertisement by reducing its length. The advertising manager is glad to retain a long-standing customer and the offer to put the extra insertion in a vacant space at an unspecified date provides flexibility. There are always cancellations and it does not cost the journal any extra to fill such spaces with repeats of regular advertisements.

Source   Private communication.

Example   **4.3(c)**

Subject   **Negotiation failure caused by a third party**

Background   A group of engineers from a developing country went to Europe to attend a specially designed programme on port management. This 18 week programme was arranged by an academic institution in cooperation with the managers of four representative ports, and included attachment periods of two to three weeks each at three of them.

To enhance his contribution, one port manager decided to arrange an afternoon for the visiting engineers with the well-known partnership of consulting engineers that had been much involved in the design of the port facilities and also in project management during their development. The budget for the total programme was very tight and in order to keep expenses down and have money available for other activities he managed to negotiate an agreement in principle that the charge for this visit would be approximately one-third of the usual commercial rate. Such special deals are not unusual, but his success was partly due to his good relationship with the partnership.

Unknown to him, however, one of the academic organizers of the programme also had contacts within the same partnership. Quite independently, the academic sought to arrange a two-day attachment for the visitors with this partnership and offered a per-day rate fairly close to the full commercial cost. Thus, when the port manager returned to formalize the arrangement already negotiated he was informed that the consultants would be pleased to accept the visitors for a two-day attachment at almost the commercial rate! This also meant that the remainder of his programme for them had to be altered in order to accommodate the new arrangement.

Comments   A lack of understanding of negotiation steps, coupled with poor communication between the organizer and the manager, undermined the latter's negotiating position with the consulting engineers. In effect, his endeavours to gain a good deal for the visiting engineers became wasted effort. This, unfortunately, is not an uncommon situation and yet it is one that could often be prevented by means of a quick phone call or short letter.

Source   Private communication.

4.4 PROJECT MANAGEMENT

Goal

To ensure, by careful coordination of the various activities involved, that all the objectives of a project are met.

About the subject

A project, particularly one undertaken by a business organization, is a specific assignment which has as its objective the task of completing contract requirements according to the specifications, on time and within the agreed resources.

This can be a real challenge, particularly with complex engineering projects involving various contractors and large numbers of items. Typical projects would be the construction of a processing plant, the installation of an offshore platform for oil production, or the building of a motorway. The number and variety of tasks in these projects and their interdependence make it difficult to estimate accurately the resources needed and the work content for each stage. This in turn affects both planning and prediction of performance. Both are necessary, however, for success in work of this type, with its tight time constraints and demand for high productivity.

As a consequence of these factors it has become accepted practice to form a project team and appoint a project manager to coordinate the whole operation. At the same time, with increased experience, various planning techniques have been developed to aid efficiency.

The principal role of a project manager is to achieve balanced coordination of three key aspects of the project:

- *Resources*:   Both human and financial resources have to be utilized in the most effective way.
- *Time*:   The schedule must be accurately planned, and carefully implemented throughout the period allotted for the project.
- *Technology*:   The most appropriate types of technology should be selected for the work involved in the project.

Key questions and issues

The principal phases of an engineering project

In order to manage an engineering project efficiently it is important to understand what is involved in each phase of a project. There are basically six phases (see Table 4.4), and the activities involved in each are as follows:

- *Concept phase*:   A number of concepts are generated to meet the specifications defined by the client. Not all those proposed at this stage are necessarily feasible from the commercial or technical points of view, but they need to be considered to provide stimulation so that a potentially good option is not dismissed too early.

Table 4.4   The principal phases of an engineering project

PHASE	ACTIVITY
1	Concept
2	Pre-engineering
3	Engineering
4	Procurement
5	Construction
6	Commissioning

- *Pre-engineering phase*: This phase is often called 'Preliminary Design' or 'Feasibility Study'. The aim is to design the preferred concept and assess its economic possibilities so that its technical and commercial feasibility can be established. The results of this phase will determine whether the selected option will proceed to the next stage or whether an alternative will have to be adopted.
- *Engineering phase*: It is during this phase that detailed design information is determined (and the term 'detailed' is often employed to indicate that numerous calculations are done using accurate methods). In addition, full constructional information is amassed, detailed planning is done and costs are carefully worked out.
- *Procurement phase*: The task during this phase is to make available—at the right time and place—all the material and equipment needed for the project, at a price within the budget. The items involved may be very complex or very routine, but failure to meet the three criteria could lead to serious delays in completion of the project.
- *Construction phase*: This phase is concerned with translating the specifications, design details, plans and procured materials into the planned physical product. It can involve either manufacture from raw materials, or the assembly of pre-formed components to produce the finished structure.
- *Commissioning phase*: The task here is to ensure that everything is working according to contract specifications prior to handover, and that certificates of approval or their equivalent have been granted. The project team's responsibility ends when the handover has taken place.

The tools for project management

A large number of techniques have been developed and refined for practical application since the project management concept was first used in connection with the building of submarines in the United States in the 1950s. Four of the most popular techniques are highlighted here:

- *The network*: A network consists of nodes and straight lines respectively representing events and activities, together with the direction taken by the latter. Analysis of a network will identify the critical paths or series of activities requiring the greatest time allotment and ways can then be found to reduce this. Networks are simple to construct and helpful for checking the logic of an agreed series of steps, but it takes skill and experience to derive the greatest benefit from their use. See Fig. 4.1 for an example.
- *The bar chart*: Here a project is broken down into a logical sequence of manageable tasks. These activities are laid out in rows while the columns represent relevant time units and the chart also indicates the time required to complete each task. The main advantages of this technique are its simplicity and the broad picture it provides, but it is not geared to a concurrent study of cost factors. See Fig. 4.2 for an example.

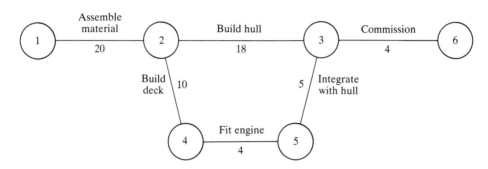

Figure 4.1 A simplified example of a network

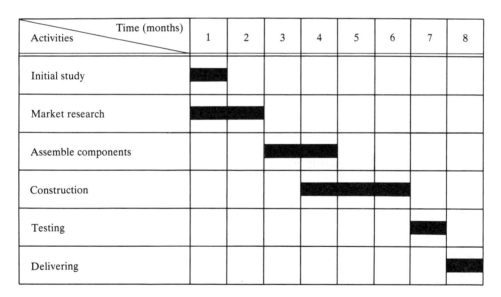

Figure 4.2 Planning with a bar chart

- *The S-curve*: When cumulative expenditure is plotted against time the shape of the resulting curve will be that of an elongated letter 'S'. S-curves are used to compare planned and actual expenditure and can illustrate this effectively, but they do not lend themselves to detailed analysis. See Fig. 4.3 for an example.

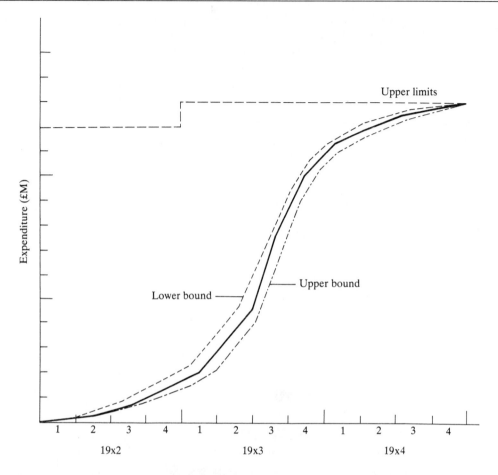

Figure 4.3   An S-curve to monitor expenditure against time

In certain projects the S-curve consists of two components representing the two boundaries of lower and upper expenditure values as also shown in Fig. 4.3. So long as actual expenditure remains within these two boundaries progress is considered acceptable. The main advantage of this approach is to avoid the project team spending valuable time on subjective assessment of how closely the actual expenditure agrees with the planned curve!

● *Critical path methods*:   For these methods a computer is essential. They are sophisticated versions of the network, but with probabilities assigned to each activity, e.g. the shortest, likeliest and longest times. They can make more realistic information available, and allow time and resources analyses to be performed, but accurate results depend greatly on the quality of past data.

Project control

Firm control over the use of both human and financial resources is necessary if a project is to be completed satisfactorily on time. Careful monitoring is necessary of the number of people employed and the range of skills in use, for example, and expenditure needs to be phased so as to maintain an acceptable cash flow, Effective project control is dependent on a number of interrelated factors, but two that are crucial are compatibility and information flow.

- *Compatibility*: Unless a control system has been built into the organization from its inception, the one that is adopted must be compatible with its existing systems. Areas that should be vetted carefully are the reporting network and the structure of the company's hierarchy.
- *Information flow*: If a project is to be properly coordinated every single member of the workforce needs some information about what is to be done, how, and by when, and those with greater responsibility will require to know much more. It is therefore vital to have a clear-cut procedure for information flow.

The role of a project manager

Staff are required to implement project management, and the key position is that of project manager, whose role is somewhat similar to that of an orchestral conductor. The responsibilties of this position include:

- *Fulfilling the management's objectives*: The project manager needs to have a clear understanding of the management's overall objectives and those for the specific project in hand. His performance will be measured by the degree to which they have been achieved.
- *Coordination*: Every aspect of the project has to be coordinated, and it is essential to ensure that relevant information is passed to the appropriate people both inside and outside the organization.
- *Financial control*: Resources have to be so controlled that income and rate of expenditure are in harmony.
- *Contract administration*: The project manager handles all matters connected with the project contract, and it is useful for people in this situation to have had experience of negotiation.
- *External links*: The project manager has to deal with clients and subcontractors, so must be able to act with firmness and yet be diplomatic at all times.

Composition of a project team

A number of people are required to support the work of the project manager, because most engineering projects are highly complex. The composition of the project team will, of course, be determined by the type, size and duration of the project in question, but some of the following will be found in any typical project team:

- *Project engineer*: Ensures general coordination between project and other personnel.
- *Construction manager*: Responsible for monitoring the construction work.
- *Quality assurance manager*: Must check that quality assurance requirements are being satisfied in every design, procurement and construction activity.
- *Service manager*: Responsible for providing information for monitoring and control purposes and for providing services to support the work of the project.

The attributes of a good project manager

What good personal attributes does a project manager require? 'As many as possible', is the answer, but—more realistically—the following are particularly desirable:

- *Objectivity*: The ability to see the total picture at all times without becoming involved in too much detail or losing sight of the ultimate goal.
- *Communication skills*: Verbal and written, with the capacity to deal with people of different languages and cultural backgrounds.
- *Alertness*: Awareness of the environment in which the work is being done and sensitivity to the possible effects of politics, religion and customs.
- *Leadership*: The ability to lead and motivate team members to achieve the necessary level of output.

ILLUSTRATIVE EXAMPLES

Example	**4.4(a)**
Subject	**Success and failure in major projects**
Background	In order to assess whether any project—large or small—is a complete success, a partial success or a failure, one has to consider how far the initial objective has been fulfilled. In general, it is not easy to gain access to the data for making such a judgement but the book by Morris and Hough entitled *The anatomy of major projects* provides insights from eight such cases. Four engineering cases will be considered here from the point of view of their project management. They are:

- *Concorde*: The supersonic aircraft developed and constructed jointly by the British and French aircraft industries with the support of both governments.
- *The Advanced Passenger Train*: British Rail's project to develop a high-speed train for use on existing tracks.
- *The Thames Barrier*: The construction of a barrier to protect London from flooding.
- *The Fulmar North Sea Oilfield*: Shell's offshore production installation using a tanker as storage and offloading system.

For measuring performance from a project management point of view the key criteria are:

- *Quality*: How closely have the specifications been met?
- *Time*: Is the work up to schedule?
- *Cost*: Is expenditure within the budget?

Application of these criteria to the above four projects yielded the results shown in Table 4.5.

Table 4.5   Success or failure from engineering projects

Project	Criteria			Status
	Quality	**Time**	**Cost**	**Status**
Concorde	Technical success	Delayed 7½ years	700% of original budget	In operation
Advanced Passenger Train	Specifications not met	Delay of 14–30 months	Within budget	Cancelled
Thames Barrier	Technical success	Delay of 3¼ years	400% of original budget	In operation
Fulmar	Technical success	Delay of 10 months	Within budget	In operation

A number of factors can contribute to a failure to meet these three criteria. The key ones are:

- Poor organization and a lack of firm project leadership
- Ineffective budgeting with no allowance for escalation or inflation
- Lack of commitment by senior management
- Poor control of project tasks
- Inception of the programme before decisions on technical aspects are finalized

Comments   Major projects are complex matters and the factors contributing to their progress are not always those connected with the management of the projects themselves. The project, for example, may be a politically motivated one. Even in such cases, however, effective application of good project management techniques will take a project a long way towards satisfying the criteria for success.

Source   Morris, P.W.G. and Hough, G.H., *The anatomy of major projects*, John Wiley, Chichester, 1987.

Example   **4.4(b)**

Subject   **Managing a major offshore oil production project**

Background   In June 1978 Britoil (now part of British Petroleum), working on behalf of the Shell and Esso oil companies, discovered oil in the North Sea at a location 290 kilometres south-east of Aberdeen in Scotland. After appraisal and delineation the site was found to be commercial, and was named the Clyde Field on 5 June 1980. In December 1982, the development plan for this joint venture was approved. Construction began in 1984, leading to the installation of the 90m tall steel jacket structure in July 1985.

Oil production began in March 1987, some six months ahead of schedule. This was a major achievement in project management with significant financial implications for the partners in the venture. Many factors contributed to this success, and a number of special features deserve attention:

- *Awareness of the overall objective*:   The project manager made sure that nobody involved in the project ever forgot its overall objective, i.e. to ensure that oil production began on schedule. This was typified by the setting up of a 'day counter' in the office, next to the model of the Clyde jacket structure, to show the number of days to oil-flow. Staff entering the office would see the changes in situation each day!
- *Ensuring efficiency and team spirit*:   The 500 strong design team, consisting of core staff and groups from various contractors, worked under one roof in a former car plant near Glasgow Airport. Locating the design office on an 'isolated' site led to increased efficiency through lack of distraction. People worked longer periods of effective hours than normal to complete assignments. Good communication was also possible between colleagues as there were plenty of opportunities to meet and discuss progress.
- *Professional respect*:   The Clyde platform involved the construction of many components from steel frames to facilities on the deck, such as the power generator and accommodation block. The total value of business was in the region of £92 million. The approach adopted in dealing with

contractors was directed at ensuring that they were confident of the professionalism and fairness of the management team. This is illustrated by the way in which contractors for different tasks were selected without the use of auction techniques. The proposals that met the closing date for submission would be opened in the presence of all eligible parties and a choice would be made on the basis of the information contained therein.

Comments   A major project such as this one is highly complex and involves a large number of activities. Delays in any one of the more critical activities will cause progress to slip back. In offshore work, the installation has to be done in the summer months and delays could put this stage back by as much as twelve months. The completion of topside deck facilities is also dependent on weather conditions. To achieve success it is vital to ensure that everyone has the right attitude to the task.

Source   *The Clyde Project*, Britoil PLC, Glasgow, 1987.
Private communications.

Example   **4.4(c)**

Subject   **The role of a project manager**

Background   This example is taken from the mechanical engineering company mentioned in Example 4.2(c), and deals with the follow-up to the winning of a contract. The potential project manager is brought into the discussions when the tender has reached the questionnaire phase and also takes part in the negotiation of the contract. Appointment to the post of project manager as such takes place only after the contract has been awarded.

In this case, the project was valued at around £40 million, and the company was committed to the design, construction and erection of a major installation in Europe. The project manager had an engineering degree and worked in the company's Technical Department before joining the Project Management Division as a project engineer. Having acquired experience over a period of four years, he now faced his first assignment as a project manager.

After studying the requirements of the project in detail he attended the 'Start Up' meeting chaired by the commercial manager.

This meeting brought together the expertise of the following:

- Project engineer
- Procurement manager

- Quality assurance manager
- Planning manager
- Accounts manager
- Design manager

Its purpose was twofold. Firstly, it would allow the commercial manager to brief those who had not been involved in the bidding process. Secondly, it would instigate the calculation cost and effort breakdown for the complete job.

The project manager then took over responsibility for meeting the objective of the project. This was achieved by means of careful coordination to ensure satisfaction of the specifications and completion on time and within the budget. The specific tasks involved were carried out by personnel assigned to the contract and comprised:

- Design
- Procurement
- Construction
- Installation
- Commissioning

Throughout the work period the project manager held regular meetings and discussions to ensure that targets were being met and expenditure was within the limits of the S-curve. He also reported to the clients on progress.

Comment   This project manager was not involved until the final stages of the contract bidding phase, but once appointed he had the responsibility of coordinating the expertise of the various specialists, and giving particular attention to cost control.

Source   Private communication.

4.5 INFORMATION MANAGEMENT

Goal

To have every item of information and data so organized that it is readily available for reference and practical application.

About the subject

The broad term 'information' is used for any collection of facts, figures, knowledge, relationships, experience, decisions or assumptions. 'Data' normally implies information as well, but it is sometimes used when the information is assembled in a structured or numerical form.

In order to function properly, an organization depends on a pool of relevant and up-to-date information. It draws on this, together with past experience in devising policies, forecasting, decision making, and the production of required goods and services. Well-managed information, in fact, is as much of a business asset as financial capital. However, the comprehensive nature of information and the amount available may make it difficult to handle effectively. For example, a shortage makes it difficult to draw conclusions, whereas too much can be equally unhelpful if the necessary facts cannot be accessed as soon as required or their correlation with other data is uncertain. The collection of data will only create storage problems without providing any real benefit unless prior plans are made for processing and application.

The efficient management of information and data is thus a vital contribution to the effectiveness of any organization, and particularly a business enterprise.

Key questions and issues

Principal methods of handling information

The two principal methods of handling information are the databank and the database, and their key features are as follows:

- *Databank*:   The items of information are stored in standard formats, and accessed either manually or by means of a computer. No direct relationships link the different sets, such as, for example, the accounts of the various companies managed by one bank.
- *Database*:   The items of information are stored electronically in a logical manner and linked to each other by predetermined relationships. They have to be searched for and accessed by means of a computer. For example, a company's database might link the product cost information from a number of suppliers under one heading and could provide a hierarchical listing of their range of products.

In order to get the best out of the system in either case, careful thought must be given to the proposed uses of the information.

A procedure for information management

The information requirements of different organizations depend on the types of activity they undertake and their application, and thus it would be difficult to lay down precise recommendations for information management in any one case. Careful examination of the systems available, however, suggests that it is possible to generalize a procedure of six key steps. Table 4.6 summarizes these steps, and the following points should be noted:

- *Define*:   It is necessary to define the type of information needed, and to quanitfy how it is to be applied in different situations.
- *Classify*:   It is important to classify items of information in a logical and systematic way so that their relative importance can be established.
- *Organize*:   The collection and storage of information and access to it should be organized as efficiently as possible. The cost of the system, its compatibility with systems already in use and scope for future expansion all need to be considered.
- *Assign*:   Specific people must be assigned responsibility for management of the information database in order to ensure the quality and availability of its contents.
- *Encourage*:   All members of the organization should be encouraged to make use of the facility.
- *Maintain*:   The usefulness of the information available can only be maintained by regular checking, updating and upgrading.

Table 4.6   A procedure for handling information

STEP	ACTION	TASK
1	Define	information needs
2	Classify	information in a logical manner
3	Organize	collection, storage and access
4	Assign	managerial responsibility to individuals
5	Encourage	usage by everyone
6	Maintain	usefulness by regular updating

The contributions of the computer

Since the amount of gathered information can increase very rapidly to the point where manual management is highly inefficient, the accepted way forward is to use computer databases. A great many types of computer software are now available for the management of information, suitable for both large computing systems and personal computers. With associated graphics packages the information can often be provided in visual forms such as spreadsheets, charts and graphs.

Aspects of computer usage needing attention

In the management of information for an organization it is important to put someone in charge, and to give special attention to:

- *Compatibility*: All hardware and software used within an organization should be fully compatible, unless there are very good reasons for some deviation. It is both unhelpful and uneconomic to have hardward in different departments that is not capable of accepting common software. A policy of standardization should be vigorously pursued.
- *Software vetting*: The selected software should do exactly what is required and not simply approximate to this. It is also necessary to ensure that the introduction of a computer for a particular task will reduce and not—as can happen—increase the time spent on it without any real gain.
- *Training*: Staff should be given proper training on the management and use of information systems.
- *Linking*: The maximum usefulness will be derived from a management database if it can be linked to sources of technical information such as the computer-aided design (CAD) database.

How important is maintenance?

The need for efficient and effective maintenance of a database cannot be over-emphasized. No database can make its full contribution to the success of an organization unless accuracy is regularly checked, errors are eliminated and data are regularly updated. Marketing decisions based on out-of-date market information, for example, are unlikely to be sound. In order to obtain value for the money invested in a computerized information system the management of an organization must also be prepared to allot adequate resources to, for example, the employment of suitably qualified staff to manage and maintain the database.

Right of access to stored information

In building up a computer database or similar system, great care must be taken to avoid infringement of national law. In the UK, for example, there is resistance to the storing of personal details on a database. In most cases the data must be accessible to the people concerned. Rights of access to data vary between countries. Access is more open in the USA than in most European countries, but, equally, in the USA the quantities of stored data on individuals are vast. Management must be alert to potential problems in this area, and ensure that any use made of personal or other information is appropriate and acceptable.

ILLUSTRATIVE EXAMPLES

Example	**4.5(a)**
Subject	**Computer-aided management of key information**
Background	The successful operation of any company depends on up-to-date and accurate information, and this is particularly important for large companies with several divisions each operating their own speciality but all involved in a similar area of activity. This example focuses on a mechanical engineering company with major activities that include the following:

- General engineering
- Pipeline work
- Offshore and petrochemical work
- Defence projects

Although every division has used computers for information management for some time it was only in the mid-1980s that it was decided to integrate and develop the information on selected areas of activity into general databases which could be accessed via workstations in every division.

One such database was concerned with selling and tendering. Its objectives were to facilitate the following:

- Coordinating sales effort
- Monitoring tendering status
- Checking on contract progress

A team, comprising staff from the marketing department and representatives of the in-house computer systems department, was formed to define the requirements for this database and to specify the features to be available in the software. The outcome was a system which offered twenty possible types of report generated from four basic sets of information. The information on 'existing and potential customers', for example, included the following:

- Name of company and contact data
- Activity group to which it belongs
- General data on its business activities
- Names of contacts within the company, matched to individual sales representatives
- Visit reports and prospect of contracts
- Status of any tendering
- Possible or necessary further actions

By being able to gain ready access to this information the company was able to organize its activities more effectively and also serve the customer more efficiently. Typical benefits of the system included:

- *Avoidance of confusion*: Prevented several members of staff getting involved in the same or similar tendering enquiries.
- *Visit coordination*: Reduced the possibility of sales representatives from different divisions travelling to see the same customers, especially in other countries.
- *More efficient tendering*: The database allowed tendering managers to keep track of the various tenders in process, with regard to follow-up needs, submission deadline for the bid etc.
- *Effective customer service*: In the absence of the responsible member of staff, the database enabled a deputy to deal efficiently with a customer's needs.

Comments   Once the users' full requirements had been defined it was possible to prepare the software in stages so as to meet an increasing range of needs. When these information management systems became available at all the workstations, the efficiency and effectiveness of the company's operations were greatly increased.

Source   Private communication.

Example   **4.5(b)**

Subject   **Computer database on companies**

Background   Developments in database management have now reached the stage where business activities can benefit from their application. One example of the type of service on offer is the provision of data on all the publicly listed companies.

Highlights of the annual accounts and other information on these companies is stored in a computer database, and the companies themselves are usually grouped according to industrial sectors, e.g. Airport Equipment. The data can be directly accessed by clients via any compatible computer terminal.

The information available may cover a number of years, and typical items would include:

- Company profile
- Analysis of profit and loss account
- Balance sheet analysis
- Ratios of key measurements of performance

- Comparison with other companies in the same industrial sector

The main benefits to users of this service can be summarized as follows:

- Ready access to company information without the need for a physical search
- Easy assessment of a company's performance, and comparison with others in the same sector
- Possibility of close competitor monitoring

Many public libraries keep a record of company accounts, but a physical search can be time-consuming, and the information on a particular company may not be up to date. The payment of a modest sum to a computerized information service will usually bring forth the necessary data.

Comments   Calling on the services of a computerized information service is the most effective way of obtaining data about commercial enterprises with the minimum of time and effort.

Source   Press comments and private communications.

Example   **4.5(c)**

Subject   **Information about the offshore business**

Background   Companies must have the relevant information if they are to be able to exploit market opportunities, or—perhaps more important—they must know how and where this information can be found. For the gas and oil industries it is the Offshore Supplies Office (OSO) that provides a vital information service.

The OSO was set up by the UK Department of Energy with the objective of promoting the UK offshore supplies industry to gas and oil operations throughout the world. It has achieved this by ensuring that its team of experts maintains close links with both operators and suppliers, and actively encourages research and development in the new offshore technologies.

Amongst OSO's many activities is the provision of relevant information to all interested parties. It regularly publishes booklets about aspects of its work, and two of these are particularly relevant in the present case.

One is the report on *Information offshore*, key sections of which are entitled:

- About the activities of the OSO
- Outline of Britain's offshore supplies industry

There then follows a series of lists of the names and addresses and the telephone and telex numbers of various organizations, under four main headings, as follows:

- *UK Government advice and help*:   175 contact points for government departments and divisions
- *UK Government research and testing services*:   22 contact organizations
- *British organizations supporting the offshore industry*:   A list of 80 names, in seven subsections, e.g. Trade associations; Industrial research associations
- *Oil companies licensed for UK oil and gas operations*:   Addresses of offices in London, Aberdeen and elsewhere in the UK for 77 offshore and 43 onshore companies

The second booklet is entitled *Oil and gas international purchasing contacts* and gives contact details for each organization listed. The list is divided into seven major sections, one for each region of the world, i.e. Africa, Australasia, Europe, Far East, Middle East, North America, South America. The information in each section is provided in three subsections, covering 'Operators' procurement contacts', 'Major indigenous and state oil companies' and 'UK commercial attachés'.

Comments   These two booklets together can save hours of valuable time, as they provide a very comprehensive range of information on the oil and gas industries which is not easily obtainable elsewhere. They would be a good starting point for newcomers to the offshore industry, and can also provide fresh contacts for those already involved, for example operators seeking expertise in a particular area.

Source   Booklets published annually by the UK Department of Energy through its Offshore Supplies Office.

4.6 RESEARCH AND DEVELOPMENT

Goal

To bring existing products or services more fully into line with the specifications by overcoming identified problems, to improve on their current performance, and to increase the organization's range by turning novel ideas into new products or services.

About the subject

'Research' is the term normally used to describe study or investigation directed at increasing knowledge and understanding of an idea, concept, problem or phenomenon. The immediate applicability of the results is a consideration, but not the principal aim since increased understanding of the particular issue should, in the long run, lead to a new product, or at least to the improvement of an existing one. 'Development', on the other hand, takes the most commercially promising research results on to the stage of practical implementation so that, for example, a new design or product can be put on the market in due course.

The borderline between 'research' and 'development' is not always clear in the techno-logical industries, and responsibility for the two activities is often vested in a single department or division. The composite term 'research and development' (R&D) is used for the complete process of increasing knowledge about an idea and solving any prob-lems identified right through to the stage of practical implementation.

It is important, however, to have a clear understanding of the differences between 'research' and 'development'. The key ones are:

- Development usually begins after research has been completed.
- The cost of the development phase is usually several times greater than that of the research phase.
- The results to be obtained from undertaking a research project cannot always be predicted accurately beforehand. This is especially true of projects involving the introduction of new techniques or approaches. Development, however, takes a more predictable path.
- Research is expected to employ new approaches, whereas development work usu-ally concentrates on extending available technologies.

Key questions and issues

Why be concerned about R&D?

Organizations should take research and development seriously because it is through improvements and the generation of new products and services that they keep ahead of competitors. However, there are two major reasons why a company may be reluctant to invest money in R&D.

- Predicting the cost of an R&D project accurately is difficult because of the inevitable 'unknown' factors, and the decision makers may be reluctant to commit funds and effort to open-ended projects.

- There is no guarantee that investment in a specific project will necessarily lead to eventual profit.

It is important to take a positive attitude to R&D and to use suitable criteria for determining the company's own level of research commitment.

Basic steps in research

Once an organization has decided in principle to undertake research, whether by means of in-house studies or by commissioning work from external institutions, the procedure involved can be broken down into six basic steps, under the following headings:

- *Identify*:   The first task must be to identify areas requiring research investigation and to establish, by means of a critical review, their current status.
- *Select*:   The approach to a given research task depends on the type of problem involved, but the major ones are: theoretical studies, computer simulation, laboratory experiments and full-scale measurements, or a combination of these four approaches. The choice in each case must be based on the fullest possible information to ensure that the most promising approach is selected.
- *Decide*:   If it were possible to define precisely what a research programme will involve there would generally be no need for the research! In view of this fact, the level of resource commitment to be accorded to any particular research project should be carefully considered. Thereafter it is the researchers' responsibility to organize their programme within the agreed time-scale and budget.
- *Do*:   Once the approach work programme and budget have been agreed, the research itself has to be carried out within the allotted time. Progress can be monitored by means of regular meetings and reports.
- *Acquire*:   The outcome of any research is usually a better understanding of the idea, concept or problem in question. There is usually an increase in knowledge about such matters as what is feasible in a short time-scale and what would take a lot longer. New problems deserving investigation are also likely to be identified.
- *Outline*:   Once a research task is completed it is helpful to identify aspects of the findings that could have practical application, and also to consider whether there are other areas in which the results of this work might make a useful contribution.

Completion of the six research steps will lead onto one of several possible situations. These include:

- The need for further work on a particlar issue
- The option of developing the results into a product
- The failure of the research undertaken to provide a satisfactory solution to the problem in question

The complete procedure is summarized in Table 4.7 and an understanding of it will help to make effective decisions about research projects possible.

Table 4.7   A research procedure

STEP	ACTION	TASK
1	Identify	areas of interest and current status
2	Select	possible approaches
3	Decide	the resources commitment level
4	Do	the work as defined
5	Acquire	understanding and knowledge
6	Outline	application possibilities

Basic steps in development

Development work usually follows completion of a research project, and attention is normally then focused on solving specific problems which have been identified through a detailed consideration of the planned product and its cost and other implications. Development work may also be undertaken to overcome identified weaknesses in the performance, appeal or cost-effectiveness of an existing product and thus improve its marketability. In either case, the basic steps involved can be classified under the following broad headings:

● *Examine*: The first task is to examine the details of the project, in particular the design implications, costs, value for money and available technologies, so as to decide on the way forward.
● *Quantify*: Specific tasks requiring attention will have been identified following the studies in the previous step and decisions must now be taken on the preferred solutions for these, taking into account the technology and equipment available, its reliability, and the costs involved.
● *Determine*:   A decision has to be made on the amount of resources the organization is willing to commit to the project. This will involve time and human and financial resources.
● *Perform*:   The necessary work is done to overcome the identified practical problems. The solutions offered must be such as can be readily implemented by available technologies.
● *Make*:   In order to achieve the desired improvement the normal practice is to introduce modifications and refinements to the design of the product or system.
● *Prepare*:   Once the development tasks are completed, consideration must be given to the ways in which the results can be put into practice. Typical questons that arise

include: 'How would this improvement be made and tested in practice?', 'How would it affect the environment?', and 'Where can it be tested experimentally?'.

It is usual to make a decision regarding implementation on completion of the development phase. Factors to be taken into account include: cost versus benefits; justification of the work based on the results of market research; the time-scale involved; how current manufacturing practice will be affected by implementing the resulting modifications; and staff training implications.

The procedure is summarized in Table 4.8.

Table 4.8   A development procedure

STEP	ACTION	TASK
1	Examine	research project details and ways forward
2	Quantify	problems requiring solutions
3	Determine	the resources commitment level
4	Perform	the task as defined
5	Make	design modifications and refinements as required
6	Prepare	for the next stage

The level of R&D commitment

The amount of R&D a company undertakes depends to a large extent on the type of business activity and the degree of competition faced. Most companies today do set aside specific sums for R&D. They are usually calculated as a percentage of turnover or profit, but the percentages can vary widely. High technology organizations in, for example, the computer industry may well have an R&D budget representing 5–12 per cent of their turnover in order to keep ahead in a fiercely competitive market, whereas heavy engineering companies with established products and a more traditional approach may limit their R&D to around 1 per cent of their turnover. R&D can be done in-house, via contractors or jointly with other companies. Each approach has its own cost and technological advantages and disadvantages.

Sources of ideas for R&D

Matters requiring R&D attention may stem from original invention, competitive pressure, market research data, customer complaints or an individual's lateral thinking. The source of a new idea may be a member of the company's own staff, an external organization, a university department, a research institution or a member of the general public. There is never any shortage of problems to overcome and ideas to consider, but the really good ideas are scarce. Because of this it is essential for R&D decision makers to have a sound procedure for filtering out the really good from the apparently good so that investment can be concentrated on those with commercial potential.

Establishing R&D priorities

Commercial organizations always have more problems requiring research attention and ideas for development than money to pay for the work, especially large organizations with an in-house R&D unit. There is a real need for effective methods of prioritizing conflicting claims for expenditure.

It is usual to separate R&D requirements into short- and long-term categories. The former generally receives the priority because solutions are needed immediately before improvements can be made or unexpected problems have arisen. The R&D effort must be concentrated on finding viable solutions as quickly as possible. Longer-term research priorities are much more difficult to determine and a range of factors has to be taken into consideration, e.g. the market potential of the product or service, the amount of time and level of funds required, the experience of competitors etc.

ILLUSTRATIVE EXAMPLES

Example	**4.6(a)**
Subject	**Project selection: technological or commercial criteria?**
Background	In the late 1960s and 1970s two main types of safety glass were used for car windscreens. Each had advantages and drawbacks, but the key issue was the type of injury experienced by front seat occupants when they came into sudden contact with the screen. Toughened monolyth glass was highly resistant to impact but caused serious head injuries. On the other hand, the laminated windscreen broke readily on impact and thus prevented head injuries but caused facial laceration.

At that time about 27 000 motor car injuries per year in the UK were attributed to glass, the majority arising from sudden contact with toughened glass windscreens. The Triplex Safety Glass Company, a subsidiary of Pilkington Brothers Ltd, was given the task in 1970 of developing a glass which could reduce or eliminate the risk of facial laceration on impact while retaining the good features of the laminated glass. This was a major technological challenge. A successful outcome would offer attractive market potential because the trend was towards adopting laminated windscreens rather than of toughened windscreens.

The R&D assignment, known as the Ten Twenty project, had a multi-million pound budget and employed many highly qualified engineers and scientists. After almost eight years they were successful in overcoming many technological problems and meeting many demanding production requirements, such as the capability for providing reliable glass in large quantities. The technological achievements received international acclaim and honours such as a Queen's Award for technical innovation and a Design Council Award. The research team also gained the prestigious MacRobert Award for scientific advances.

A number of car makers had specified Ten Twenty windscreens for their cars and it was expected that new legislation on windscreens would also provide an impetus for the introduction of this new product. It was therefore decided to invest in a multi-million pound new factory which came into operation in 1978.

A combination of unfavourable factors, however, prevented the Ten Twenty glass from becoming a commercial success. Typical factors included:

● Defensive measures taken by car manufacturers who had their own glass-producing capabilities and capacities
● Decline in the UK car industry

The most important determinant was the passing of the UK legislation to make compulsory the wearing of seat-belts by the occupants of front seats in cars. When this became law fewer head injuries were caused as a result of windscreen impact. This led the board of Triplex to take the hard decision to mothball the Ten Twenty factory soon after its commissioning, with heavy financial losses. On the positive side, however, the developed process technology did allow Triplex to win aircraft glazing business against strong international competition.

Comments   The commercial failure was attributed to the fact that a law which required all new cars to be fitted with laminated glass car windscreens was not passed in the UK. However, the real cause of failure was the fact that the wrong question was asked by the management. The emphasis was placed on whether Triplex was capable of producing a 'super' windscreen glass. The answer was a firm 'Yes'.

The commercial question that should have been asked was: 'Can the glass-making capability of Triplex win a significant market share in the mass transportation business?' The answer would have been in the negative because it would have to face competition not only from other glass manufacturers but also from alternative methods of increasing car safety and preventing injury to occupants. The key 'competitor' was the legislation that required every front seat occupant to wear a seat-belt. This greatly reduced the risk of injury due to windscreen impact and meant that the superglass would only find a market if priced lower than alternative types of windscreen.

Source   Nicolson, R., *Research and development—examples of success and failure*, Philips Lecture, The Royal Society, July 1988.

Example   **4.6(b)**

Subject   **Methods of selecting project for R&D**

Background   Companies employ various methods to determine the topics to consider and the level of funding to commit to research and development work, and three typical approaches are highlighted here.

● *An engineering services company*: A small company enjoys a good demand for its services, which range from design studies to the training of clients' staff in the use of facilities and the operation of installations. The technical requirements of a particular task are always identified before the contract is signed, but problems can arise in the course of the work. The managing director will then set in motion an R&D project in order to find a solution. Depending on its nature, the work may be done either in-house or by a research organization.

● *A major manufacturing company*: This company manufactures a range of glass products for markets worldwide. It has a research department staffed with highly qualified engineers and scientists and their terms of reference are to solve problems and thus develop new products for the company. Each year the research staff draws up a list of ideas, and the Research Director's team decides which areas to fund for the coming months. The Director of R&D has a crucial influence on the decisions about research and the allocation of funding.

The Director of R&D during the 1950s and 1960s was responsible for a major breakthrough in the mass production of quality glass. His two successors, keen to match up to their famous predecessor, have each expended considerable research effort and resources while endeavouring to make a further similar breakthrough.

● *The engineering department of an oil company*: A group of five people have the responsibility for coordinating the department's R&D needs. By means of structured discussion with colleagues this group identifies problems requiring solution. These are then prioritized and requests for funds are put to the Resources Committee of the Engineering Department. These requests usually have to be modfied before an agreed figure is incorporated into the budget for the coming year, as the amount available is usually relatively small. The allotted funds may be used internally or spent on contracts with consultants, research organizations and university departments.

Comments There is no definitive procedure for selecting R&D topics. In industry, the emphasis is generally on seeking solutions to problems encountered in the manufacture or use of products. Industrial companies engaging in fundamental research aimed at a major breakthrough, are likely to experience failure and disappointment unless careful consideration is given to the commercial factors involved.

Sources Private communications and press comments.

Example **4.6(c)**

Subject **Sources of research and development results**

Background Companies depend on research and development in order to maintain progress, whether this implies increasintg its share of the market, arresting a decline in sales or improving the efficiency of its manufacturing.

Traditionally, R&D results can be obtained from the following main sources:

● Professional journals, and reports by researchers in the available published technical literature
● Reports by in-house R&D units within companies
● Reports on commissioned projects from institutes or laboratories that specialize in research or development

In an article in the *Financial Times*, David Fishlock discussed the findings of a working party set up by the European Industrial Research Management Association (EIRMA), a club or more than 200 European companies actively involved in research. The objective of the study was to determine how these companies could acquire advanced technology in the most effective way for the development of new products and processes.

The working party studied two types of industry, and their findings were as follows:

● Type A: Assemblers of products using components and sub-systems made elsewhere. Examples of these include manufacturers of aircraft, motor cars, and ships. The R&D results in this case were often very sophisticated and most of them were generated by the suppliers of components (or on their behalf by another organization). The key approach here is 'design and integrate'.

- Type B: Manufacturers of a complete product. Examples would be producers of bulk chemicals and glass-makers. Research is mainly done in-house. The key approach is 'design and make'.

Both approaches were found to be effective, but as products and processes become more complicated it is essential to encourage the Type A approach in which in-house R&D will be only one part of the total activity.

Comments   With the cost of R&D being so high it makes a lot of sense to use results from every available source. Assembling companies or industries have to decide carefully between the core technologies on which to concentrate in-house R&D, and work to be left to their suppliers. The real problem arises when a 'missing area' for research is identified. Should the company take on this task itself or encourage a supplier to do so? The correct decision in each case must depend upon an overall assessment of the situation.

Source   Fishlock, D., The European Industrial Research Management Association, *Financial Times*, 5 July 1990.

4.7 MATERIAL FOR FURTHER STUDY

Ashford, J.L., *The management of quality in construction*, E. & F.N. Spon, London, 1989.,

Dean, B.V. (ed.), *Project management: methods and studies*, Elsevier, Barking, 1985.

Fulmer, R.M., *The new management*, Macmillan, London, 1988.

Goodman, L.J., *Project planning and management*, Van Nostrand Reinhold, London, 1988.

Hajeh, V.G., *Management of engineering projects*, McGraw-Hill, Maidenhead, 1984.

Kennedy, G., Benson, J. and McMillan, J., *Managing negotiation*, Hutchinson, London, 1987.

Lochyer, K. and Gordon, J., *Critical path analysis and other network analysis technologies*, Pitman, London, 1991.

Lock, D. (ed.), *Project management handbook*, Gower, Aldershot, 1987.

Oakland, J.S., *Total quality management*, Heinemann, London, 1989.

Oldcorn, R., *The management of business*, Pan, London, 1987.

Scott, W. and Billing, B., *Negotiating skills in engineering and construction*, Thomas Telford, London, 1990.

Wearne, S.H., *Civil engineering contracts*, Thomas Telford, London, 1989.

Welsh, A.N., *The skills of management*, Wildwood House, Aldershot, 1989.

CHAPTER 5

MONEY

Money, the subject selected for this chapter, is examined from various points of view. These range from the influence money has on the performance of an organization to the way in which money is affected by economic variations and governmental decisions, and include possible sources of funds.

The first section deals with the fundamentals of economics and begins with a definition of this topic. Aspects examined include demand and supply, international trade, the balance of payments, interest rates and employment levels.

Four principal sources of funds are considered: commercial organizations, government bodies, international agencies and special organizations. The key merits and drawbacks of each source are highlighted.

Accounting methods are then introduced, and a range of topics is considered with two purposes in mind. The first is to assist readers in acquiring some skill in understanding company accounts. The second aim is to help them learn how to obtain useful guidance from the data.

The performance of an organization from a financial point of view is dealt with in the next section. The key information available is discussed first, followed by an outline of suitable methods for assessing performance and productivity, and for making comparisons between companies. This is followed by a section devoted to assessing the

117

viability of project proposals. Three key methods are highlighted, these being the Payback, Net Present Value and Internal Rate of Return methods.

The sixth section examines the effects of government policies with respect to both national and local governments. This topic is brought under the heading of 'Money' because many actions resulting from changes in policy have direct or indirect financial implications.

The final section contains a list of material for further reading.

5.1 UNDERSTANDING ECONOMICS

Goal

To provide an appreciation of the process of generating and distributing wealth and of the direct and indirect influence of economic issues on the performance of a business organization.

About the subject

Economics is the study of the process that attempts to balance human aspiration—which can be unlimited—and finite resources. The former is expressed as demand for products, services and finance, and the latter is represented by land, raw materials, production facilities and human skills. The importance of this subject is reflected in the fact that our daily lives are constantly affected by economic factors, for example, the number of jobs available.

Economics is a complex subject with a large number of variables which themselves are dependent on many imponderables. It is usually treated as a subject with two principal components, known as microeconomics and macroeconomics. Microeconomics deals with issues relating to specific industries and markets and how they affect organizations and individuals. Typical examples include such matters as:

- The demand for goods and services
- Prices and pricing
- Investment decisions
- Financial evaluations and assessments

Macroeconomics, on the other hand, examines issues on a global scale, and considers such aspects as:

- International trade
- The level of taxation
- Industrial expansion and recession
- The supply and management of money

Economics involves making forecasts of future trends in these areas to enable appropriate decisions to be made and plans to be devised. The information is used, for example, by governments in allocating resources and planning for the future. This is an extremely challenging task and typical problems relate to the accuracy of forecasts, the making and implementing of decisions, and examination of their effects. The outcome of such decisions will generally extend over a lengthy period of time and each stage can be affected by many factors, such as changes in government. An appreciation of the fundamental issues of economics is essential if an organization is to perform well in a constantly changing economic environment. Most of the attention in this section will be focused on macroeconomics, because many aspects of microeconomics are considered in other sections.

Key questions and issues

The economic approaches available

Business activities are affected by the economic approaches adopted by governments together with other closely related factors, such as the cultural background of the country, the local availability of natural resources and its industrial capability.

In general, a government's economic philosophy can lie anywhere between what is known as the *laissez-faire* or extremely free approach and the 'planned' or highly centralized approach.

In the *laissez-faire* approach, organizations and industries are allowed to compete freely with each other in response to customer requirements. In this situation only the most efficient can expect to survive, and there is limited government intervention when difficulties arise. With the centralized approach, on the other hand, the government dictates the entire policy and makes devisions regarding what will be produced and the price at which products will be sold. However, in times of difficulty government support can also be expected.

A tool for the study of economics

As in engineering in economics investigation is often done by means of a technique known as 'theoretical modelling', or simply, 'modelling'. In this case a real economic system is represented by a mathematical formulation or model and its behaviour under a variety of conditions is then simulated by means of a computer. The effect on the system of varying the values of different parameters singly or in groups can then be gauged. This procedure can isolate the parameters which are most significant in the economic situation, and can also provide some indication of future trends. It should be noted that the accuracy and usefulness of the results will depend on the assumptions used in the model, the way it is formulated and the quantity and accuracy of the data used.

The basic characteristic of economics

To understand the economic performance of an entity such as a country, it is necessary to examine its output or some other aspect over a long period of time, such as 5, 10 or even 20 years. The salient feature of an 'output versus time' graph is the fluctuation between peaks and troughs, which is similar to the responses of a dynamic system plotted against time. The behaviour of this curve or cycle provides relevant insight into the performance of the entity in question.

There are four basic phases in the cycle. The expansion phase for a country is when the gross domestic product (GDP) is increasing. The recession phase is when real GDP decreases. The peak and the trough are the highest and lowest points in economic activity. The terms 'boom' and 'depression' are used respectively for prolonged periods of expansion and recession. It is very important for a business organization to develop the capability to anticipate the peaks and troughs, as well as the probable duration of expansion and recession phases. Both periods offer opportunities, but they have to be exploited in completely different ways, in the former case, for example, by expanding facilities and in the latter by diversifying. In general, government policies tend to be directed at minimizing the impact during the fluctuation so that inflation may be dampened and slackness removed by active stimulation.

What determines the employment level?

A high level of employment is usually associated with a period of industrial expansion and a low level of employment (or high unemployment) is associated with recession. This is readily seen because in a growth period, for example, in order to meet the increased demand more people are needed to provide services or to generate products. The opposite is the case during a downturn of activity. Unfortunely, the real life situation is far more complicated, because of the number of other variables involved. Typical of these are changes in production methods, political decision etc. Economic theories provide possible explanations for various levels of employment (and possible solutions) to enable their occurrence to be predicted.

How are demand and price related?

In an ideal competitive market, the price of a product or a service would be determined solely by the relationship between demand and supply. In practice, the situation is far less straightforward. In the first place, a large organization may choose to sell products at a special price for a strategic period in order to gain a market share and thereby cause distortion in the competition. Secondly, local shortages can provide an opportunity for alternative suppliers to come on the scene, creating extra parameters which have to be taken into consideration. Thirdly, there is the effect of possible government action in the form of credit restriction, taxation or new regulations.

The importance of international trade

National prosperity from year to year is governed to a large extent by business exchanges between nations, or international trade. For example, one country may be supplying raw materials while the other may sell manufactured goods. In an ideal situation, there should be a balance of trade, i.e. what a country buys and sells should be approximately equal in value. In this way all parties will benefit from the strength of each, and effective use is made of scarce resources. International business, therefore, not only offers greater scope for expansion of market opportunities, but can also increase national well-being.

Factors influencing the balance of trade

Firstly, a trade balance can only be achieved if the goods involved are of a standard and price that will attract customers. When there are other suppliers with similar goods, there is bound to be strong competition to ensure that one supplier's, or one country's, goods are selected in preference to those of another. Bargaining between governments may temporarily alter the balance of trade, but there is no guarantee of longer-term success in selling in a particular country.

Secondly, one country may not have sufficient resources to pay for another's goods or services, as demonstrated by the fact that raw materials often have a low value while products manufactured from them can have great added value. During the past 15 years, for example, the prices of raw materials have altered only marginally while the prices of manufactured goods have increased steeply. This imbalance of purchasing power affects trade patterns between nations. In time, most countries would, where possible, wish to produce the majority of their basic requirements themselves, or even to export these products so that less money is spent on foreign imports.

Thirdly, some form of protection is exercised by most countries to ensure that their products gain a 'fair' share of world trade. Politically, trading methods can be a very sensitive issue. For example, if one country puts import restrictions on another's goods the other may take retaliatory action which could have a detrimental effect on the trade of both nations.

The effect of exchange rates

Money is used as a medium for trading purposes in order to avoid bartering with actual goods. Since most countries have their own currency, during trade transactions the need arises either to exchange currencies, for example, the pound sterling with the yen, or to use a common currency acceptable to all trading parties, such as the US dollar. The performance or economic status of each country, as measured by such factors as its trade earnings, can affect the value of its currency, which means that there is a continual fluctuation in exchange rates.

Exchange rates can have a major influence on the profitability of a business. Favourable exchange rates will help to sell goods and services abroad, leading to greater profitability through a high volume of trade, but the reverse is also true.

In order to reduce the fluctuation in exchange rate between the currencies of its members and hence assist in stabilizing their business activities, the European Community has developed an Exchange Rate Mechanism (ERM) for linking together the currencies of member nations.

The nation's standard of living

Trading has a key influence on a country's wealth and hence on the standard of living of its people. A country which manages to achieve trade surpluses year after year will eventually enjoy a higher standard of living, as represented by higher incomes. The ability of a nation to compete successfully in trade can be increased by investing in modern production facilities and training people in new skills so as to increase efficiency. This is turn puts pressure on the ability of other nations and their business organizations to compete effectively.

ILLUSTRATIVE EXAMPLES

Example   **5.1(a)**

Subject   **The allocation of resources**

Background   One of the key issues of economics is the allocation of resources. This is a matter that concerns the individual, as well as the commercial venture and the government. Whether they are plentiful or very limited it is always a challenging task to ensure that resources are utilized effectively. Examples of approaches in two different situations are given here.

The UK Government's planned spending for 1989–90:   The pie chart in Fig. 5.1 shows the UK Government's planned allocation for major areas of spending during the period. It should be noted that about one-half of the total is allocated to social security and the health and social services. Defence expenditure takes up about one-eighth of the total.

The distribution alters from year to year, but the scope for redistribution in the short term is quite limited. In practice, whenever a change is proposed each sector will try to put forward a strong case for an increased share of the budget. Defence, for example, will insist that the country is

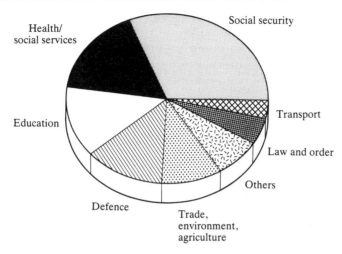

Figure 5.1   The UK Government's spending plan for 1989–90

going to be 'too exposed' unless expenditure on it is maintained at a 'realistic' level. Equally, cuts in the education budget will be hotly contested, with evidence of how other successful nations spend much more in this area than the UK does.

The spending plan of a postgraduate student: This example is based on the planned allocation of funds prepared by an Indian postgraduate student who has won a scholarship to study in the UK. He expects his outlay to be distributed as follows:

- Accommodation: 35 per cent
- Food: 33 per cent
- Travel: 5 per cent
- Clothing, footwear etc: 6 per cent
- Books, stationery: 8 per cent
- Social: 7 per cent
- Miscellaneous: 6 per cent

It can be seen that this student allowed around two-thirds of the total amount for accommodation and food. He would have liked to spend more on books and other reading materials but has little room for manoeuvre within the budget as all the other items are either essential or are themselves underfunded.

Comments   The total amounts involved in these two examples are very different, but the difficulty in apportioning them to the best advantage is the same. There is always a conflict over which items deserve the most funds. Ideally the 'unit of benefit' or value for money in each case should be the same as a result of the allocation.

Source   Private communications and press comments.

Example   **5.1(b)**

Subject   **Attracting foreign investments**

Background   Many countries seek to attract foreign investments as a method of increasing the standard of living or the economic welfare of the nation. Typical approaches include: inviting foreign companies to explore for minerals on land or in coastal waters, in the hope of developing a processing industry with export potential; and seeking a transfer of technical know-how through encouraging foreign manufacturers to set up local subsidiaries to make a selected range of products.

The factors which will determine the attractiveness or otherwise of a country to investors are numerous. The principal ones are national policies, the political stability of the nation and its economic potential. The example considered here is based on material provided by the Malaysian Government for business people interested in investing in their country.

A one-page information sheet describes the country, its economy and some basic regulations, and this is backed up by two pages of statistics and a diagram showing the major economic indicators for the period 1985–88. A summary of key facts for 1988 is as follows:

- Area of country: 329 758 km^2
- Population: 16.9 million
- Per capita income: M$4815 (US$1 = M$2.71)
- Real growth of domestic product; 8.1 per cent
- Unemployment: 8.1 per cent
- Leading economic sectors: Manufacturing, agriculture, mining
- Manufacturing products: Electrical and electronic products, textiles, clothing, rubber-based products
- Commodity products: Palm oil, natural rubber, tropical timber, tin, cocoa beans, pepper

Various incentives are available to stimulate investment, and examples include:

- The 1986 Promotion of Investment Act provides tax incentives for foreign investors, an export allowance, and incentives for research and development etc.
- Agreements have been signed with a number of countries to protect

foreign investors from having to pay tax twice on international income.

- Investment guarantee agreements safeguard the interests of foreign investors.
- Foreign exchange arrangements allow funds to be freely transferred in and out of Malaysia.

Over the period in question the balance of trade has been positive, with linear increments between 1986 and 1988.

Comments   Foreign investment is now a highly competitive business. Many countries are therefore, now setting up advisory offices in strategic locations abroad in order to assist potential investors.

Source   Press comments.

Example   **5.1(c)**

Subject   **Assessing economic performance**

Background   There are many ways of assessing the economic performance of a country, for example, by measuring trends and productivity and the output of manufactured goods. Two popular methods are described here.

Balance of payments: A measure of a country's economic strength can be obtained from the trends in the value of its balance of payments, i.e. the difference between the values of its exports and its imports. Strong economic performance is indicated by a positive figure, or a high level of exports, while a deficit is indicated by a negative figure or a high level of imports. Four countries have been selected for this example, for the period from 1970 to 1988:

- The United States of America
- The United Kingdom
- Japan
- The German Federal Republic

Figure 5.2 shows how the four countries compare. It is not surprising to find that Japan and Germany had strong economies and the most impressive records for the period in question. Trade deficits were, however, the main feature of the USA and the UK's performance over the same period.

Real rate of interest: The 'real rate of interest' is defined as the difference between the base rate of interest and the inflation rate. Its effects can be best illustrated in the form of a graph (see Fig. 5.3), which shows the real rate of interest for the UK economy between 1952 and 1988 together

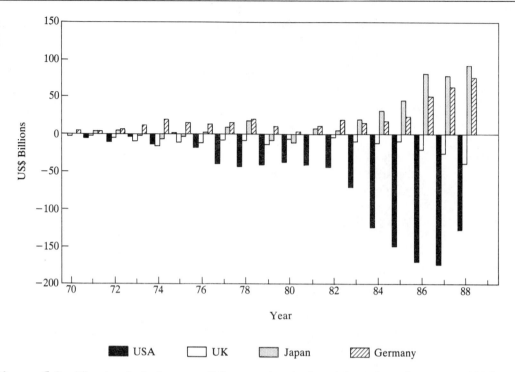

Figure 5.2   The trade balances of four major industrial nations between 1970 and 1988

with the curves of the base rate of interest and the inflation rate. For reasons of clarity, the inflation rate is plotted as a negative item.

There are several points to note:

- For the period between 1953 and 1969 the real rates of interest were positive, although the average value was only 2.0 per cent.
- During 1973 the price of oil increased threefold, leading to the so-called 'first oil crisis'. In the course of the next two years the real rate of interest became negative, i.e. money lost its value at an alarming rate, and in 1975 its value was down to − 13.8 per cent.
- In the 1980s the real rate of interest returned to a positive value, although this fluctuated. The average rate for the period 1980 to 1988 was 5.1 per cent.

Comments   It is essential to examine a range of factors in order to obtain a complete picture of a nation's economic performance. These factors include: return on investment, stability of government, free trade policies, rate of inflation,

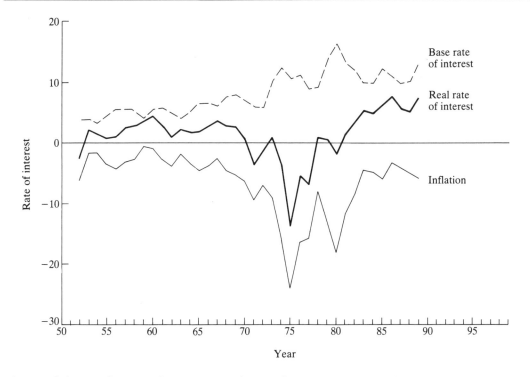

Figure 5.3   Real rates of interest in the UK between 1950 and 1970

and balance of payment. This example illustrates the last two factors in the list.

Sources   Press comments and private communications.

5.2 SOURCES OF FUNDS

Goal

To identify where financial resources can be found to support a business activity at various stages of its development.

About the subject

Most organizations require financial assistance at every stage of their development, but the amounts will vary with the types of activity undertaken at any one time. Additional funding is particularly necessary, for example, when an organization is starting up, when the product range is being expanded, or when a decision is taken to diversify. Money can usually be found somewhere for a proposition with sound commercial viability or potential, but it is often difficult to find propositions that will readily meet this requirement. It is useful to have some knowledge of the various sources of funds and the types of venture they can support, together with their relevant merits and drawbacks. There is a very large number of sources of funding, which can be broadly divided into four main groups:

- Commercial sources
- Government bodies
- International bodies
- Other sources

As can be seen from Fig. 5.4, there are within each group a number of options and care needs to be exercised to ensure their eligibility and suitability. Some of these sources will make funds available to organizations that are likely to share any profit with shareholders, while others will assist only so-called 'non-profit-making' organizations which have no shareholders.

Key questions and issues

Types of commercial sources

These sources can be broadly classified as follows:

- *Banks*: Clearing banks and merchant banks that specialize in funding major projects.
- *Venture capital companies*: Organizations set up to provide capital and assistance to those wishing to create companies to exploit specific opportunities or to expand into fresh areas.
- *Investment trusts*: Money is invested in these funds, and is then lent out by them for business activities.
- *The public at large*: Funding is obtained from this source through sale of shares to intending investors.
- *Pension funds*: These funds may make money available for long-term projects.
- *Insurance companies*: Again, funds are made available for long-term activities.
- *Public sector agencies*: Government-backed organizations which specialize in giving selective support to business activities.

What can banks offer?

Banks, for example, will provide individuals or companies with fixed term loans for specific projects. They are also prepared to give business concerns an ongoing overdraft facility, which may run to several million pounds in the case of a large company. In fact,

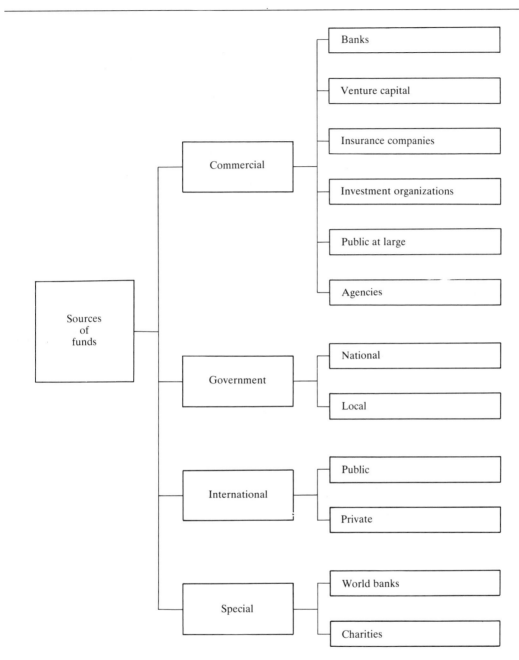

Figure 5.4 Possible sources of funds

the bulk of any bank's income is generated by lending money to such borrowers at a higher rate of interest than they offer to savers. The main merits of this source are that money is available for lending purposes, and in most countries there are several major banks and each has a large number of branches in different cities and locations. However, banks tend to favour 'safe' business ventures.

Venture capital companies

This is a method that works well in the USA. Groups of organizations put money into a 'venture capital' company with the specific aim of investing funds in new ventures to exploit business opportunities. Such groups understand the need for adequate capital in setting up a new business venture and are prepared to offer this, and can also provide expert guidance to help ensure that it will succeed. They do, however, look for an early return on capital invested, and are more likely to be attracted by ventures with prospects of a quick return. Also, their contribution usually takes the form of share capital, which gives them a strong say in the running of the company. Despite this, venture capital companies are a good source of finance, particularly for new business ventures.

The general public as a source of finance

Public companies in many countries can obtain development finance by offering shares for sale to the general public through the Stock Exchange or its national equivalent. This is the classical way for an established business organization to raise capital. Large amounts of funding can be obtained by this means and there is virtually no restriction on its use. However, a company wishing to trade on the Stock Exchange has to have 'good credibility', or a record of financial integrity. This is usually indicated by a good performance record and good future prospects.

The contribution of government bodies

Both national and local governments and their agencies can provide financial assistance for certain activities, usually under strict conditions. This funding can take various forms, e.g. grants, loans, subsidies or tax concessions. Certain government departments may provide commercial organizations with funding for a relevant research and development project of up to 50 per cent of the total costs involved. The UK Department of Energy, for example, has made various awards for hydrocarbon research and development. Such funds are available for special initiatives, where it can be shown that the results will assist the fulfilment of government plans in the targeted area. However, their usage is often restricted and work may have to be shared with competitors.

The role of special government agencies

In many countries, funds are available for business initiatives from the local or regional government or from semi-government agencies, such as the Canadian Provincial and US State authorities, and the UK's Scottish Development Agency. This last was set up to support industrial and business developments in Scotland alone. It uses money provided by central government, but this is administered at the local level. Its main advantage is that the staff have specialized knowledge of their own sphere of responsibility and are keen to support local initiatives. However, the conditions imposed on such support may not always be sufficiently flexible.

What can international sources offer?

A variety of international bodies administer funds which may be available to organizations for research and development purposes although applications usually have to be carefully processed through a number of stages. A typical example of such a body is the European Community (EC). Member states pay annual contributions to the community and money from the central fund is then disbursed in a number of different ways. As well as funding for projects in specified 'developing regions' it encourages transnational research and development work with potential for future business activity which could benefit the whole community. For most such schemes, it is mandatory that groups in at least two countries must be involved in the work and the EC provides only a percentage of the total funding needed. A typical scheme is ESPRIT, for advances in information technology.

The contribution of the World Bank

The World Bank was set up by the United Nations in order to use finance from its 'richer' members for the express purpose of helping to improve the standard of living in developing countries. Through the governments of these countries funds can be obtained for business developments. Typical schemes include an education and training scheme for Indonesia, and a scheme for improving water transportation in Bangladesh.

Other sources of support

Funding may also be obtained from a number of other sources for organizations which have 'non-profit-making' status. Many large organizations and wealthy individuals, for example, put part of their annual profits into a foundation from which deserving causes and certain types of business project may be funded. Examples include the Ford, Wolfson and Kennedy Foundations. Their money is usually provided as a grant and repayment is not required. However, since the amount available for annual distribution is limited, eligibility for awards is usually restricted in various ways.

ILLUSTRATIVE EXAMPLES

Example   **5.2(a)**

Subject   **Funds from a bank**

Background   The most popular source of borrowed funds for engineering activities is a bank, and each major bank has its own procedures and criteria for lending money. Success in securing a loan usually depends on a combination of factors, the four key ones being: the commercial potential of the proposed venture; the credibility of those seeking the loan; the present performance of the business (unless the venture is a new one); the availability or otherwise of some form of security.

The approach to lending which has been selected as an example is that employed by the Clydesdale Bank, one of the three major banks in Scotland.

Prior to any discussion of a loan the Clydesdale Bank usually recommends that the intending borrower reads its own set of four booklets on business matters. Written in an easy-to-follow and reader-friendly style, they are entitled:

- *Checklist before going into business*
- *First steps in setting up a business*
- *The business plan*
- *Monitoring and control of your business*

These booklets were prepared primarily to assist aspiring entrepreneurs in answering a questionnaire which provides the bank with essential information on the potential borrower and the planned venture.

By far the most important factor for assessing the funding worth of a project is the borrower's business plan. This plan provides the bank with the information on the following: relevant experience of the borrower; the background of the project; proposed products or services; markets; financial considerations and projections; cashflow forecast and financial requirements. It is on the basis of this information that judgement is made regarding whether to make a loan, how much to lend, for how long, and at what rate of interest.

The following are some statistics on business lending by the Clydesdale Bank:

- Around three-quarters of applicants are successful.
- The duration of a loan can vary from three months to over ten years.
- The interest rate charged depends on whether the loan is unsecured or secured against some asset, e.g. land or other property belonging to the borrower.
- In general, the rates for secured and unsecured loans are 2 per cent and 5 per cent respectively over the base rate (the current standard rate laid down by the Bank of England).
- The precise form of any loan depends on individual requirements. Some borrowers ask for an overdraft facility, while others have immediate need of a specific sum which will be repaid over an agreed period.

Comments Banks require to be satisfied with regard to a number of key factors before they will grant business loans. In this regard, the approach of the Clydesdale Bank to potential borrowers can do much to ensure that both parties have a sound appreciation of business activities and of the market potential of the proposed product or service.

Sources *First steps in setting up a business*, *Checklist before going into business* and *Business plan*. Booklets from the Clydesdale Bank.

Example **5.2(b)**

Subject **Guidance on sources of business funds**

Background Knowing where to find the money for a given purpose is very important to a business, and this was recognized by the Bank of England when it joined forces with the City Communications Centre to produce a book entitled *Money for business*. It was first published in 1978, and has since reached its fifth edition. Although certain aspects of financing may have changed over the years, this book is still useful both as a guide and as a valuable source of information.

In the introduction the Governor of the Bank of England stresses a point which is often emphasized by those with business experience, i.e. that funds are almost always available for good projects. The important thing is to know where to find them. Many managers are not aware of the various sources they can tap nor which are the most appropriate for a specific need. *Money for business* seeks to dispel this ignorance.

The book contains three parts, dealing respectively with Business finance, sources of finance, and Contact information. Part 1 contains sections on:

- How to assess financial needs
- Introduction of new equity capital
- Borrowing

These are followed by six illustrative examples. This part closes with some good tips on matters about which a financing agency usually wants information.

Part 2 contains sections on:

- Equity capital
- Short-term finance
- Medium-term finance
- Long-term finance
- Finance for exports

- Public sector finance and assistance
- Finance from European sources

The initial emphasis is on identifying the type of finance required and its proposed use. The different facilities are then briefly described and key features highlighted. Each section also contains notes on the main sources of the specific type of finance under review.

Part 3 consists of a comprehensive list of addresses and telephone numbers of relevant organizations. Information on all the banks in England, for example, is given under the heading of 'Clearing banks and other domestic deposit banks'. The 60 pages of this part include data on:

- Clearing banks and other domestic deposit banks
- Other banks and lending institutions
- Discount houses
- Credit insurance companies
- Investment trust companies
- Pension funds
- Stock exchange
- Export houses
- Venture capital institutions

Comments   All those involved in management would be well advised to devote some time to studying this short text even if finance is not their direct responsibility. The knowledge gained will do much to assist their company in achieving its objectives.

Source   *Money for business*, 5th edn, Bank of England and City Communication Centre, 1985.

Example   **5.2(c)**

Subject   **Funding from a charitable foundation**

Background   The Wolfson Foundation makes a significant sum available to educational institutions in the UK every three or four years for research and development activities. To initiate each new round of funding the Trustees select a particular theme and invite applications for grants for relevant projects. For example, in 1987, the theme was medical projects. The objective was, however, the same as that of previous schemes, i.e. the grants would be for research projects which could lead to the 'development of a product or process for use' in medicine or the medical industry. In general, the Trustees tend to favour projects involving more than one discipline, for example, physical science or engineering departments working with medicine. Applications also have to show that there is a 'need' for the potential product or service and to indicate how the results of the research are to be exploited. If such exploitation generates income for the institution, then the Trustees would ask for a small amount of that income to be transferred annually to the Foundation for a fixed period in order to support future initiatives.

The actual format of the application form provides some indication of the thinking behind the Trustees' requirements. The form is in two parts. Part I asks for the basic data on the applicant, together with a statement of the objective of the proposed work and an abstract. The lengths of these must be limited to 30 and 250 words respectively.

Part II asks for the following additional information:

- Title of project
- Description of proposed project, within a maximum of 1500 words
- Curricula vitae and track record of applicants
- Financial statements, including: funds required, cashflow forecasts, selected projects, proposed method of sharing any income generated
- Contact data for three referees

Since this is a very popular way for academics to try to raise 'business' finance, the Foundation's invitations tend to attract a large number of applicants. Some of the proposals put forward are very good, but others are unlikely to meet the specifications. A certain amount of pre-selection is therefore done by the submitting institutions, who can send forward to the Foundation only three applications each in any one year.

There is a high degree of competition for these awards and the judging of the proposal is consequently extremely strict. There is also an approximate ceiling to the amount provided for a single project, as the Foundation likes to spread the funds available over a reasonably large number of proposals.

Comments   Hopeful applicants will find that their scientific case is greatly enhanced by an ability to write clear and concise objectives and attractive abstracts, together with a sharp focusing of the commercial potential of the project.

Source   The Wolfson Foundation letter to the academic institutions, September 1987.

5.3 INTERPRETING ACCOUNTING INFORMATION

Goal

To provide information on the financial aspects of an organization in forms that allow its performance to be readily interpreted and its future potential assessed.

About the subject

For an organization to operate efficiently it must have ready access to three basic items of information. First and foremost it must be able to check the cashflow position, or know whether there is sufficient income or funds available at any given amount to cover day-to-day operating costs, loan repayments, the purchase of equipment and, in the case of a company, shareholders' dividends. Secondly, it needs to know how profitable its operations have been over a given period, such as one year. Lastly, it is necessary to know what its future potential is. Some of this information is in fact a legal requirement in most countries. Companies registered in the UK, for example, must send a copy of their annual accounts to the Registrar of Companies.

Accounting, in business terms, is the compilation and provision of basic information in a 'standard' format for the management to consult and compare and to make future projections. The same information is useful to engineers in certain aspects of their work. Yet it is surprising how large a percentage of engineers find business accounts confusing, although the processes involved are so much simpler than the mathematics they need to use in practice!

Bookkeeping is the accurate recording of every income or expenditure transaction, and it will provide much of the required information. The old manually entered ledgers are now giving way to computerized records, and software is available to provide an instantaneous review of the current financial status of the organization. Its effectiveness, however, depends on having accurate data fed into the computer punctually and systematically. Bookkeeping is also the first step in the determination of an organization's performance. The key relations linking income (I) and expenditure (E) are as follows:

Profit:	$(I - E)$ is greater than zero
Break-even:	$(I - E)$ is equal to zero
Loss:	$(I - E)$ is less than zero

In practice, of course, the matter is rather more complex, because of the variety of ways in which income is received and expenditure is incurred.

Key questions and issues

The importance of cashflow

Cashflow is the difference between income and expenditure, both of which alter with time. It can be represented by the following mathematical relation:

$$\text{Cashflow} = \int_{t=t_0}^{t=t_i} [I(t)-E(t)]\,\mathrm{d}t$$

where $I(t)$ and $E(t)$ are income and expenditure at a time t. The cashflow at any time is the sum of the differences between these items between one time t_0 and another time t_1. In other words it is the integral of the difference. It is important because it indicates the viability of a business. When cashflow is positive, or profits are being made, fresh initiatives, expansion and consolidation can be undertaken. Negative cashflow, or deficit, can be tolerated for only a limited period of time before operation ceases to be possible.

The information for measuring performance

Company law requires every business organization to provide three basic sets of information at regular intervals. These are summary accounts which are compiled from bookkeeping records for a time-span known as an 'accounting period'. This is usually one year but sometimes the results for a six-month period are also given. The three items required are shown in the Annual Report under the headings of 'Profit and Loss Account', 'Balance Sheet' and 'Sources and Application of Funds'. In each case data for the previous accounting period is presented in parallel for ease of comparison. Since these items are summaries it is normal practice to provide additional explanations in the form of 'Notes'. It should be remembered that the information provided in this way refers to the situation at a given point in time—the balance sheet, for example, is something like a snapshot. The situation could be very different at another date.

The content of the Profit and Loss Account

The general form of a Profit and Loss Account is given in Fig. 5.5. The key items are turnover, profit and earnings. 'Turnover' represents the volume of business over the period, i.e. the total amount of income generated. Profit is subdivided into operating, amount before taxation and amount after taxation, profits assigned to dividend payments to shareholders and profits retained. 'Earnings' are measured by normalization to 'Earnings per share'.

From this presentation we can obtain a useful indication of performance over the period under review and of how the results compare with those of the previous year. It is important to note, however, that the real success of an operation cannot always be

KEY ITEMS	NOTES	ACCOUNTING PERIODS	
		PRESENT	PREVIOUS
Turnover			
Operating profit	1		
Profit on ordinary activities before taxation	2		
Profit on ordinary activities after taxation	3		
Profit distribution:			
Dividends paid and proposed	4		
Retained profit of the financial year	5		
Earnings per share (pence)	6		

Figure 5.5   The general form of a Profit and Loss Account

gauged from a single year's Profit and Loss Account because it will contain information that is affected by the ways in which money is allocated for expenditure. It is usually necessary to look at an extended period for a true impression. For example, a company may invest heavily in modernizing its facilities, which causes problems of cashflow in the short-term but results in higher profits in the longer-term through increased efficiency and effectiveness.

The Balance Sheet

The general form of a Balance Sheet is shown in Fig. 5.6. It contains two main items of information. The first comes under the heading of 'Assets employed' and included fixed assets such as facilities, current assets such as cash at bank, and liabilities such as short-term borrowings. The second item comes under the heading of 'Financed by' and includes shareholders' capital and reserves and long-term loans. The total for each item is the same, and they are thus balanced. It should be noted that for a given accounting period there are two types of expenditure, known as revenue and capital expenditure, and they have to be treated differently. The former covers the current accounting period while the latter could be spread over a number of years, and this will be indicated in the Balance Sheet.

Sources and Application of Funds

The general form of this item is given in Fig. 5.7, and has three sections. The first is entitled 'Sources of funds', and there are two main sources. Funds such as operating

KEY ITEMS	NOTES	ACCOUNTING PERIODS	
		PRESENT	PREVIOUS
Assets employed			
Fixed assets	7		
Current assets	8		
Current liabilities	9		
Net current assets	10		
Total assets less current liabilities			
Creditors' amount due after more than one year	11		
Net assets	12		
Financed by			
Capital and reserves	13		
Others: loans, borrowing etc.	14		
Total funds employed			

Figure 5.6   The general form of a balance sheet

KEY ITEMS	NOTES	ACCOUNTING PERIODS	
		PRESENT	PREVIOUS
Sources of funds			
Generated from operations	15		
Funds from other sources	16		
Total			
Application of funds			
Operational expenditure	17		
Investments	18		
Interest payments, loan repayment	19		
Tax and divident payments	20		
Total			
Difference between sources/ application of funds (change in working capital)			

Figure 5.7   The general form for sources and application of funds

profits are generated internally by the operations of the company, and additional funding may be raised from external sources by means of, for example, a bank overdraft.

The second section is entitled 'Application of funds' and covers operational expenditure, investments such as the purchase of fixed assets, interest payments and the repayment of loans, the payment of dividends to shareholders and tax on profits.

The final section shows the difference between the total amounts in the previous two sections, and thus indicates the increase or decrease in the working capital of the business during the period under review.

Making future projections

So far we have been considering accounting data as they record the past performance of a business organization. However, it is also necessary to forecast probable future performance, and this forecast is usually given on a form, known as a pro forma or previously formatted sheet, similar to those used for the Profit and Loss Account and the Balance Sheet, but also including cashflow (see Fig. 5.8). The data, however, are nor-

DESCRIPTION		YEAR 1991				
	PERIOD	1	2	3	4	SUM
Income						
xxx		——	——	——	——	——
xxxxx		——	——	——	——	——
xxxx		——	——	——	——	——
	Total	——	——	——	——	——
Expenditure						
xxxxxx		——	——	——	——	——
xxx		——	——	——	——	——
xxxxxx		——	——	——	——	——
xxxxxx		——	——	——	——	——
xxxxxx		——	——	——	——	——
xxxxxx		——	——	——	——	——
	Total	——	——	——	——	——
Net inc/exp		——	——	——	——	——
Cumulative inc/exp		——	——	——	——	

Figure 5.8 A general form for a pro forma of cashflow

mally presented for several periods of, say, three or six months rather than for a single period of a year. This allows the information to be readily assessed by people with an interest in the organization's future, e.g. those considering investing in it.

ILLUSTRATIVE EXAMPLES

Example	**5.3(a)**
Subject	**Typical annual accounting for an engineering company**
Background	Although the general headings are always the same, all companies present their annual accounts in their own special style and some contain more detail than others. In fact, a certain amount of experience is needed to be able to interpret correctly the data on different organizations. Figs. 5.9, 5.10 and 5.11 show the actual data on an engineering company for the year 1988, the highlights of which are as follows:

- *Profit and Loss Account*: The key data of interest are the turnover, operating profit, profit distribution, net profit before and after taxation, and earnings per share. There has been an impressive improvement over the period in the case of all these aspects.
- *Balance Sheet*: The main sections are those dealing with assets and methods of financing. It should be noted that the figures for 'Net assets' and the 'Total funds employed' should always be the same.
- *Sources and application of funds*: There are three main sections here.

KEY ITEMS	NOTES	ACCOUNTING PERIODS	
		PRESENT	PREVIOUS
		£M	£M
Turnover		3527	2836
Operating profit	1	447	423
Profit on ordinary activities	2		
before taxation		377	393
Profit on ordinary activities	3	245	254
after taxation			
Profit distribution:			
Dividends paid and proposed	4	81	72
Retained profit of the	5		
financial year		151	173
Earnings per share (pence)	6	32.3p	34.3p

Figure 5.9   An example of a profit and loss account

KEY ITEMS	NOTES	ACCOUNTING PERIODS			
		PRESENT		PREVIOUS	
		£M	£M	£M	£M
Assets employed					
Fixed assets	7		1560		890
Current assets	8	1654		1410	
Current liabilities	9	1042		861	
Net current assets	10		612		549
					1439
Total assets less current liabilities			2172		1439
Creditors' amount due after more than one year	11		628		535
Net assets	12		1544		904
Financed by					
Capital and reserves	13		1420		794
Others: loans, borrowing etc.	14		124		110
Total funds employed			1544		904

Figure 5.10   An example of a Balance Sheet

The first gives the actual sources of funds employed, which may be generated from operations or obtained externally. The second section deals with the application of funds for internal use and for external purposes. The difference between these two items gives the amount in the third section which is entitled 'Working capital'. A surplus is shown directly in this section and any deficit is printed in brackets.

Comments   Accounts sheets may appear complicated, but all the key information can be readily found once its purpose is understood.

Source   Press comments.

Example   **5.3(b)**

Subject   **Keeping a record of accounts**

Background   The basic data needed to prepare the accounting information given in Example 5.3(a) can be drawn from bookkeeping or ledger records. These

KEY ITEMS	NOTES	ACCOUNTING PERIODS	
		PRESENT	PREVIOUS
		£M	£M
Sources of funds			
Generated from operations	15	447	423
Funds from other sources	16	61	73
Total		508	496
Application of funds			
Operational expenditure	17	330	539
Investments	18	7	5
Interest payments, loan repayment	19	70	30
Tax and dividend payments	20	201	130
Total		608	704
Difference between sources/ application of funds (change in working capital)		(100)	(208)

Figure 5.11   An example of sources and application of funds

records may be kept either in manual form or on a computer. The actual method selected will depend on the volume of data involved and the type of organization. It is important, however, to have a 'user-friendly' format for the collection of information.

The following table shows how the data can be recorded over a period of one month for a small engineering consultancy consisting of three staff.

Ref: MAX Consultants   *Period*: July 1990

DATE	DESCRIPTION	REF.	INCOME	EXPENDITURE
2/7/90	Special service income	R47	£1200.00	—
9/7/90	Purchase of stationery	R57	—	£225.00
11/7/90	Printing new brochures	R60	—	£650.25
25/7/90	Fees for senior consultants	R74	£4500.00	—
28/7/90	Salaries	R75	—	£3120.20

From this table it is possible to prepare an income and expenditure statement for the month of July 1990 as follows:

Income and expenditure statement for month: July 1990

	Description	Amount	Total
Income:	Special Income (R47)	£1200.00	
	Fees (R74)	£4500.00	£5700.00
Expenditure:	Stationery (R57)	£225.00	
	Brochures (R60)	£650.25	
	Salaries (R75)	£3120.20	£3995.45
	Profit for the month of July		£1704.55

Using the information gathered in the ledger it is then possible to prepare the Profit and Loss Account, Balance Sheet and Sources and Application of Funds.

Comments　　　There are many ways of presenting the information in a ledger but the key requirements are always clarity and the possibility of quick cross-reference. For computer data-processing it is helpful to use either standard data sheets or user-friendly screen displays.

Source　　　Private communication.

Example　　　**5.3(c)**

Subject　　　**Cash-flow projection**

Background　　　Most business plans have a section devoted to financial predictions. This example considers the cashflow projection for a small engineering consultancy in the first year of its operation. The pro forma is given in Fig. 5.12.

This projection has the following points of interest:

- There are two main sections, headed 'Income' and 'Expenditure'. 'Income' is made up of expected fees plus the borrowed loan capital, while 'Expenditure' covers salaries, rents, travel, subsistence, overheads and interest on a loan.
- The pro forma contains four columns, each representing financial activity in one quarter of the year. A fifth column shows the amount for each item over the whole year. Summing down the columns gives the total

DESCRIPTION		YEAR 1991				
	PERIOD	1	2	3	4	SUM
Income						
Fees		5 000	15 000	2 000	20 000	42 000
Loan capital		20 000	—	—	—	20 000
	Total	25 000	15 000	2 000	20 000	62 000
Expenditure						
Salaries		7 500	7 500	7 500	7 500	30 000
Rents		500	500	500	500	2 000
Travel		1 500	2 000	600	1 200	5 300
Subsistence		400	600	100	500	1 600
Overheads		4 500	4 500	4 500	4 500	18 000
Interest		750	750	750	750	3 000
	Total	15 150	15 850	13 950	14 950	59 900
Net inc/exp		9 850	(850)	(11 950)	5 050	2 100
Cumulative inc/exp		9 850	9 000	(2 950)	2 100	

Figure 5.12 An example of a pro forma of cashflow

estimated income and expenditure for each quarter. The estimated figures for the whole year are in the fifth column.

- The two bottom rows provide two items of information. The first is the difference between income and expenditure or net value, shown in the penultimate line. Bracketed figures indicate expenditure in excess of income. It will be noted that the first quarter shows a surplus of £9850 while the second quarter has a deficit of £850. The cumulative values of the quarters are given in the bottom line. In this example these are £9850 for the first quarter, £9000 (£9850 − £850) for the half year, a deficit of £2050 (£9000 − £11950) for the three-quarter period and a profit of £2100 (£5050 − £2950) for the year.

Comments   The mechanics of preparing a cashflow pro forma are very straightforward, but it is difficult to forecast accurately the figures for the individual items. It should be noted that no account is taken here of the time value of money to allow for inflation.

Source   Private communication and press comments.

5.4 PERFORMANCE ASSESSMENT

Goal

To identify the key factors which will provide an accurate measurement of the performance of a business enterprise both in absolute terms and in comparison with organizations in similar business sectors.

About the subject

Ideally, the performance of an organization should be measured from both the commercial and technological points of view, in relation to pre-defined objectives. In practice it is the commercial objectives that are used to make the assessment. There are two basic reasons for this.

Firstly, technological assessment determines how efficiently the organization is meeting the technical specifications of its product or service. However, it is not usually easy to assess this factor for other similar products or services. As a result, performance is normally assessed on the basis of commercial factors such as the selling price. For example, a poorly coordinated production line will generally lead to a higher unit cost for a product. This is particularly true when an organization is offering a large range of products or services.

Secondly, it is the commercial performance and future commercial potential that are of prime interest to existing and potential investors. Assessment information can therefore be presented in standard forms for comparison with other companies in similar business sectors.

It should be borne in mind that even in commercial assessments there are subjective elements which cannot be readily quantified in absolute terms. The performance assessment of an organization therefore calls for a combination of information which is readily available, for example in the company's annual accounts, and data on the track record of those who are running the organization.

Key questions and issues

Key information

Most countries have their own methods of presenting commercial information, but what is usually readily available can be grouped under the following headings:

- *Turnover*: This is the amount of business transacted during one accounting period, such as twelve months. It provides an indication of the size of the company and its market share if this can be clearly defined.
- *Profit*: Profit is the amount of income in excess of expenditure and this figure provides information on how well the organization is meeting its objectives.
- *Dividends*: The amount of money paid to those who have invested in the company as shareholders.
- *Share value*: The value of the company's shares and its behaviour over a given period. For publicly quoted companies this information is readily available and successful operation is usually reflected in a rate of value increase above the current rate of inflation.
- *Asset value*: The total value of the company's assets as represented by, for example, buildings, land, human resources, equipment, stocks of goods, capital reserves and materials.
- *Number of shareholders*: This information provides a broad indication of the degree of public support for the company and good performance usually tends to attract investors.
- *Number of employees*: The size of the company's human resources give a guide to the efficiency of its functioning.

The use of the key information

The absolute values of the items listed above are interesting, but they do not, in themselves, provide a true assessment of performance or a basis for comparison with other companies. A more meaningful way of presenting the data is to employ per-unit terms as ratios of one item to another, percentages and unit costs. When information is presented in this format it is much easier to compare the performance of companies in the same business sector and to compare different sectors with each other. Comparison with similar organizations in other countries is also possible.

The most popular performance data

The following figures are generally used for measuring the performance of a business organization:

- *Profit margin*: This figure is the ratio between profit and turnover expressed as a percentage. Clearly a high percentage implies good management, production efficiency, the successful introduction of new products or the effective exploitation of new opportunities.
- *Profit to capital employed*: This figure indicates the ratio of profit to the capital needed to operate over a given period. Ideally it should be as large as possible, but in general it fluctuates within a narrow range of percentages unless there have been major changes in company operations.

- *Earnings per share*:   This is the amount of profit earned in relation to each share, and it is determined by the ratio of the profit to the total number of shares issued. It is a very useful factor for comparison with other organizations.
- *Percentage increase in dividends*:   An annual rise in the level of dividend paid to shareholders is the hallmark of a good organization and will ensure the continued support of those looking for income from their shares. It should be noted that this is not easy to achieve on a regular basis.

Assessing longer-term viability

To gauge the performance of a company and its future potential it is useful to examine its records for a longer period, such as 10 years. The aspects selected for assessment purposes would include most of those given in the annual accounts. A general tabular form for their presentation is given in Fig. 5.13. The key items would be the following: turnover, profits, assets, financing of activities, and asset-worth per ordinary share.

Measuring productivity

The productivity profile of a company can be obtained from a close examination of the information provided by a company and a good guide to productivity can be obtained from the following figures:

- Turnover per employee
- Pre-tax profit per employee
- Average wage/salary per employee
- Net capital employed per employee

Such factors are useful for assessing the efficiency of other companies in similar business sectors.

Profit versus market share

Broadly speaking there are two methods of measuring an organization's performance. One school of thought uses profit as the yardstick while the other prefers to use the percentage of market share. In countries such as the USA and the UK the tendency is to assess how well a business is performing on the basis of its profits over a given period, and managements devote all their attention to increasing this figure. The argument is that shareholders will only invest in a company if they can expect dividends to rise. This is only possible if increased profits are generated. Companies in Japan and Germany, however, tend to seek an increase in the market share for their products and are less concerned with the level of short-term profits. The management in this case would claim that their approach favours the long-term welfare of the company and that shareholders will benefit from the increased capital value of their investment. The preferred yardstick in any particular case will depend on the philosophy, attitudes and government policies operating in the country concerned.

YEAR		10	9	8	7	6	5	4	3	2	1
Turnover	£M	—	—	—	—	—	—	—	—	—	—
Profit											
Profit before taxation		—	—	—	—	—	—	—	—	—	—
Earnings per share	p	—	—	—	—	—	—	—	—	—	—
Dividend	p	—	—	—	—	—	—	—	—	—	—
Assets											
Fixed assets		—	—	—	—	—	—	—	—	—	—
Current assets		—	—	—	—	—	—	—	—	—	—
less liabilities		—	—	—	—	—	—	—	—	—	—
Total		—	—	—	—	—	—	—	—	—	—
Financed by											
Share capital reserves		—	—	—	—	—	—	—	—	—	—
Borrowing		—	—	—	—	—	—	—	—	—	—
Working capital		—	—	—	—	—	—	—	—	—	—
Total		—	—	—	—	—	—	—	—	—	—
Asset worth per ordinary share		—	—	—	—	—	—	—	—	—	—

Figure 5.13 The general form of a financial summary for a ten-year period

Other key data

The information used for performance assessment, as considered so far, does not provide a total measure of its viability. To obtain an in-depth assessment of an organization's position it is necessary to look at some of the following data:

- The key people running the organization and their management track record. Typical factors to consider are management skills and strategies, foresight with respect to future development, and effective deployment of human resources.
- The organization's standing in the business sector. This is indicated by its earnings and, in the case of publicly quoted companies, by the value of its shares.
- The potential growth of that entire business sector, and the organization's strength in comparison with its major competitors.

It must be stressed that it can be very difficult to obtain such information, particularly in the case of a company that is not publicly quoted.

ILLUSTRATIVE EXAMPLES

Example	**5.4(a)**
Subject	**Information on company performance**
Background	Two UK companies have been selected to show their performance over a given accounting period:

A company with engineering and international interests

Year of Account	1988
Turnover	£7396 million
Profit	£880 million
Dividend	£268 million
Asset Value	£5349 million
Number of Shareholders	201 000
Number of Employees	105 000

A company in the building industry

Year of Account	1988
Turnover	£2836 million
Profit	£393 million
Dividend	£72 million
Asset Value	£1060 million
Number of Shareholders	45 203
Number of Employees	28 923

Comments	The absolute figures give some indication of the level of business activity of the two companies but it is difficult to make comparisons between them.
Source	Press comments.
Example	**5.4(b)**
Subject	**Performance assessment by ratios and percentages**
Background	Information about the performance of four different companies is used to illustrate the effective application of ratios and percentages.

A company in the building industry

Year of account	1988
Profit margin	14.9%
Profit to capital employed	35.8%

Earnings per share   34.3p
Percentage increase in dividend   39.0%

A shipping and construction group
Year of account   1988
Profit margin   9.4%
Profit to capital employed   8.2%
Earnings per share   53.8p
Percentage increase in dividend   16.0%

An international group with engineering interests
Year of account   1988
Profit margin   11.9%
Profit to capital employed   16.4%
Earnings per share   115.9p
Percentage increase in dividend   55.0%

A major oil company
Year of account   1988
Profit margin   8.0%
Profit to capital employed   10.6%
Earnings per share   20p
Percentage increase in dividend   8.0%

Comments	The first company benefited from building activities in the middle 1980s where very little capital was needed to generate profit. The other three companies needed a fair amount of capital in order to earn profit. The third company has been showing excellent growth for a number of years.
Source	Press comments.
Example	**5.4(c)**
Subject	**Longer term financial summary**
Background	A typical company in the building industry is used to illustrate performance assessment over a five-year period. Details are given in Table 5.1.
Comments	The results indicate a steady growth over the period in question. The outstanding feature to be noted is that the profit margin (profit to turnover) has increased during the period from 8 per cent to 13 per cent while the ratio of Profit to Capital Employed (profit to capital needed) remained between 29 per cent (1987) and 32 per cent (1984).
Source	Press comments.

Table 5.1   Financial summary for the period 1983–87

	1987	1986	1985	1984	1983
Turnover £M	2200	1735	1570	1321	1160
Profit					
Profit before taxation	265	170	135	110	90
Earnings per share p	23.5	17.8	14.1	12.9	12.1
Dividend p	7.3	5.6	4.7	4.0	3.4
Assets					
Fixed assets	765	518	461	441	285
Current assets less liabilities	242	207	143	70	52
Total	1007	725	604	511	337
Financed by					
Share capital reserves	711	492	427	334	253
Borrowing working capital	148	173	126	132	47
	149	60	51	46	37
Total	1007	725	604	511	467
Asset worth per ordinary share	99.9	79.1	69.0	54.2	46.7

5.5 PROJECT VIABILITY

Goal

To quantify the commercial viability of identified opportunities with sufficient accuracy for effective decision making.

About the subject

When an organization identifies an opportunity through market study or some other method, the first task must be to establish its viability in financial terms. This is necessary for at least three reasons.

Firstly, it is illogical to decide whether to proceed with a project solely on the basis of enthusiasm or to reject it because of the poor communication skill of its proposer. An objective decision requires critical assessment of the financial projection.

Secondly, there are usually several different possible ways of exploiting an opportunity. The long-term implications of each method need to be considered before one is selected for action.

Thirdly, most new ventures require financial support. This may be found within the organization, obtained from an outside body, or drawn from a combination of both resources.

Whatever the choice, the decision makers need to be convinced that the proposed activity has a better claim for support than other potential ventures. Support is most likely to be given according to a well-presented case based on technological ideas, which is backed up by accurate financial projections and a critical assessment of possible hurdles.

In view of these factors, it is necessary for both the proposers and those making decisions about new projects to apply a suitable procedure for assessing its viability. This is true for both the identified opportunity within the existing organization and for a new activity altogether. Such a procedure should include consideration of how the opportunity has been evolved, preparation of a business plan and appraisal of the project by a relevant set of criteria.

Key questions and issues

Assessment of viability

There are a number of ways of assessing the general viability of a project, depending on its type, the country in which it is being developed, and whether it is a new project or an enhancement of one already existing. Once the concept of a project is ready for implementation, however, by far the most logical approach for the proposers is to prepare a business plan. This will clarify their own thinking and is an important aid for those who might wish to become involved. In addition, various methods are available for assessing the financial viability of the proposals in such a plan.

What is a business plan?

A business plan is a document which systematically presents basic information about a project and its proposers, their principal activities, and the arrangements made for potentially successful implementation of the project. Where possible the information provided should include:

- The type of business organization, its background data and principal activities
- The goals and strategies of the management
- The key personnel and their background, expertise and experience
- The products or services already on offer and those now planned
- The markets for the new product or service, competitors and targeted customers
- Financial projections and cashflow forecasts
- Time-scale for reaching profitability
- Summarized plans for operation and quality control

A clear presentation of these points will help to provide a business plan that is attractive to potential supporters.

Why have a business plan?

There are several reasons why a business plan is useful, in particular, for small business ventures and activities. The key ones are:

- It helps proposers to put down on paper their ideas and solutions for the implementation of their concept. It also forces them to provide answers to questions regarding the various fundamental elements of business. Thus they are helped to 'think the project through' and draw up a working framework for key members of their organization.
- A good business plan, demonstrating clearly how a project could succeed, is of immense value to potential sponsors and backers in deciding whether or not to give it their support.
- Every project has to compete for attention with other proposals in similar or related areas and the funding is usually being provided by the same sources. A well-prepared business plan may be the crucial factor in winning a favourable decision.

Financial appraisal methods

There are two basic groups of methods for evaluating an identified opportunity. One group comes under the title of 'Payback methods'. This approach determines the point when the cumulative cash flow from the sale of the product or service will be equal to the capital expenditure required for its manufacture or provision. The other group of methods are known as 'Discounted cashflow methods'. This is essentially the reverse of the calculation used to determine compound interest, and involves working out the relative merits of different ways of employing a sum of money. For example, it allows a company to decide whether it would be better off in the long run to invest a sum such as £10 000 for, say, six years at a given rate of interest, or to spend it on developing a project with an envisaged profit by the end of the sixth year of activity. Two forms of discounted cashflow calculation deserve consideration, the Net Present Value (NPV) method and the Internal Rate of Return (IRR) method.

The payback method and its effectiveness

The payback method is the most popular one for appraising the viability of a project because it is relatively straightforward to apply. The basic items of information necessary for the calculation are usually available from the business plan, and these are:

- The amount of capital invested
- The cashflow over a given period

The cashflow over a given period is the difference between the total sales, or turnover, and the total expenditure over the period of interest. The payback period is obtained from the ratio of capital investment to cashflow. For example, an annual sales value of £100 000 and expenditure of £80 000 gives a cashflow of £20 000 per year. Suppose the capital invested was £60 000; the payback period would therefore be £60 000 divided by £20 000, which gives a period of three years.

The main drawback of this method is that it does not take into account factors such as time value of money, or unexpected changes of circumstance during and after the payback period. These features are illustrated in Example 5.5(a).

The Net Present Value method

This method is based on the concept of 'time value of money' whereby an amount of money available today is regarded as worth more than the same amount of cash earned (or cashflow) at some future date. In the case of a new product, the amount of expected profit by the end of, say, year five, has to be estimated and brought back to its value at the present time before it can be compared with the amount currently available for the project. In this way several options can be compared to reveal the one which will be the most profitable, or has the highest NPV. The key factor is the rate of interest selected for discounting purposes, bearing in mind the fluctuations that can occur over a five-year period. The basic relation is as follows:

$$(\text{Money in } N \text{ years}) = (\text{Money now}) \times (1 + \text{interest rate fraction})^N$$

Or

$$M_N = M_p(1 + i)^N$$

where M_N is the money at the Nth year, M_p is the money at present, and i is the interest rate in percentage terms. Thus the present value is expressed as:

$$M_p = M_N/(1 + i)^N,$$

and $1/(1 + i)^N$ is called 'present worth'.

For example, £10 000 in one year's time at 10 per cent interest will yield a present value of £9091, from £10 000/$(1 + 0.1)^1$.

In practice, other factors such as inflation have to be taken into account, and it is not always easy to predict the net cashflow for each period in the life of a project.

The basis of the IRR method

Using this method one calculates the discount rate so as to make net present value equal to zero, or to evaluate what discount rate will make investment worth while.

Since it is not possible to know what this discount rate will be in advance, it is usual to apply a calculation method that involves iteration. In such a case two discount rates are assumed. One will give a positive NPV and the other a negative NPV. By plotting values of the NPV against these rates it is possible to determine the discount rate at which NPV is equal to zero. This method also requires the cashflow for each year of a project's life to be estimated.

The decision on whether to proceed with a project will depend on whether the rate of return over its life is acceptable.

Contribution of cost/benefit analysis

Ideally, every potential project should be examined from the point of view of 'cost/benefit' in order to find out what benefits, in the broadest possible sense, can be derived from investing a given sum of money. In practice it is extremely difficult to carry out a realistic exercise of this nature either because sufficient information may not be available or because of a lack of the means for equating a given benefit with a specific cost. As a result it is usually only the economic information that is analysed. Yet, in the case of a new concept, it is very often not possible to justify the venture purely on economic grounds and other justifications are needed.

Essentially, a cost/benefit analysis begins by assuming the likely cost of the project and then listing the perceived potential benefits in direct and indirect categories. Typical direct benefits would be improved efficiency and lower unit cost. Indirect ones could include the creation of new job opportunities arising from the opening of additional factories.

The problems in estimating cost/benefits tend to arise over the indirect rather than the direct outcome. For example, the building of a new stretch of road to link two cities more directly should provide some at least of the following direct benefits:

- Reduced journey time for commuters
- Reduction in transportation costs
- Improved efficiency of organizations in both cities

Indirect benefits could include:

- Opportunities for building industrial centres and developing residential areas

- Reduction in the level of stress for drivers, hence improving their overall performance
- Attracting visitors to the two cities

ILLUSTRATIVE EXAMPLES

Example	**5.5(a)**
Subject	**Payback method**
Background	A new electrical machine is predicted to have commercial potential, but investigation of the project shows that the capital investment would have to be of the order of £9000. An initial estimate of cashflow yielded the following four cashflow scenarios:

YEAR	SCENARIO			
	A	**B**	**C**	**D**
0	(£9000)	(£9000)	(£9000)	(£9000)
1	£6000	£3000	£8000	£1000
2	£3000	£6000	£1000	£8000
3	£1000	£4000	£4000	£4000
4	£1000	£5000	£5000	£5000

The payback period in each case is two years, but the method gives no indication as to which one is preferable over that period. However, Scenario B is clearly more attractive than Scenario A because it offers greater cashflow potential in the third and fourth years. For Scenarios C and D further projections are the same in the next two years. The payback method is again unable to indicate which is preferable over that two-year period, although Scenario C is clearly advantageous because it will reduce borrowing more quickly. The preferred Scenario, therefore, would be C.

Comments	The payback method is readily applied to determine the payback period, but the four scenarios considered indicate clearly that it has a major weakness. By not taking into consideration the effect of time, it fails to provide a true assessment of a project's viability.
Source	Private communication.

Example	**5.5(b)**
Subject	**Application of Net Present Value method**
Background	In this example a ship owner is considering whether to invest in a second ship. It can be bought for £2 000 000 and its resale value at the end of three years is estimated to be £1 000 000. If the income from it is such that there is a positive cashflow of £400 000 per year we can evaluate the value of investing the £2M of capital by assuming a discount rate. Using 10 per cent as the discount rate, the present worth factors are determined from $1/(1+0.1)^1$, $1/(1 = 0.1)^2$ and $1/(1 = 0.1)^2$ respectively for the three-year period. The discounted cashflows are obtained from the product of cashflow and the appropriate present worth factors. The calculation can be tabulated (in thousands) as follows:

Year	Value	Cashflow	Present worth factor	Discounted cashflow
0	2000	−2000	1.000	−2000
1	−	+ 400	0.909	358
2	−	+ 400	0.826	330
3	1100	+1500	0.751	1126
		300		− 186

There is a positive cashflow of £300 000 which appears attractive. However, as the net present value is negative (at a value of £186 000), on a discount rate of 10 per cent the proposed investment should be rejected.

Comments	When the time value of money is not taken into consideration the investment appears to be attractive because there is a resale value of £1 000 000 and a positive cashflow at the end of year 3. But if the time value is taken into consideration the investment is less attractive.
Source	Press comments and private communications.
Example	**5.5(c)**
Subject	**Application of Internal Rate of Return method**
Background	Using the same basic data as in Example 5.5(b), another way of assessing project viability is to select two discount rates for the calculation so that a positive and a negative NPV can be obtained. By using linear interpolation a discount rate can be found whereby NPV = 0.

Assuming a 4 per cent discount rate the tabulated calculation is as follows:

Year	Value	Cashflow	Present worth factor	Discounted cashflow
0	2000	−2000	1.000	−2000
1	—	+ 400	0.962	385
2	—	+ 400	0.925	370
3	1100	+1500	0.889	1333
		+ 300		+ 88

A positive NPV is obtained in this case.

By linear interpolation between a discount start rate of 4 per cent yielding an NPV of £88 000 and a discount rate of 10 per cent yielding a negative NPV of −£186 000 from Example 5.5(b), we can see that an NPV of 0 will be obtained when the discount rate is 5.93 per cent. This is the internal rate of return which would make the investment worth while.

Comments   The rate of 5.93 per cent may not be very good value for money. To improve this rate of return we could do one of the following:

● Increase the earnings per year
● Reduce expenditure
● Achieve better resale value
● Negotiate a lower initial purchase price
● Apply a combination of all these factors

As an exercise, the reader should try out each of these possibilities.

Source   Press comments and private communications.

5.6 EFFECTS OF GOVERNMENT POLICIES

Goal

To quantify in commercial terms the likely impact of government decisions on an organization and, where possible, to respond positively.

About the subject

No matter which country is the home base of a business organization, its activities and prospects will be influenced in various ways by the policies not only of its own government but also those of other countries with which it has trade dealings. In addition it will be affected by those of its own regional government or local authority. The viability of an organization will be greater and its performance will be more effective if these policies are well understood by those involved. Sound strategies can then be devised to take full advantage of the opportunities they offer, and to minimize the effects of potential limitations on activity.

National government policies determine such factors as:

- Direct and indirect taxation
- The level of import and export activities
- The inflow and outflow of investment funds
- The exchange rate for the national currency
- Interest rates on borrowed money
- Grants and subsidies
- The control of pollution

Local government policies will influence the following matters:

- The use of land, via planning permission
- Local taxation
- The level of grants and subsidies for new developments
- The transport infrastructure

From a business point of view, some aspects of government policy can have a detrimental effect on particular industries but, on the other hand, government decisions can open up vast new opportunities to organizations in many different sectors of industry. Figure 5.14 shows a typical set of constraints and opportunities affecting a decision by the UK Government on permissible pollution levels from motor car engines.

The outcome of these factors is that the most satisfactory performance can be expected from companies that have clearly defined objectives, do careful market research and are able to respond quickly to changes in the operating environment.

Key questions and issues

Policies with the greatest impact

Government policy affects organizations in different ways depending on their type and their activities. One with extensive export business, for example, will be more affected

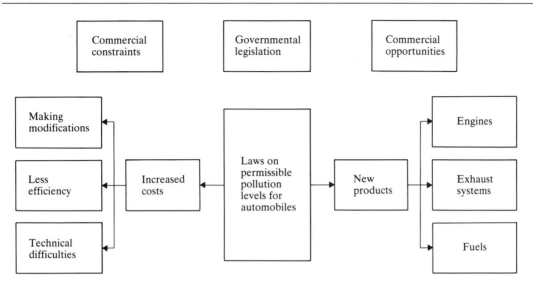

Figure 5.14 Commercial opportunities and constraints arising from a government policy

by the government's foreign policy than one with purely local involvement. The aspects of policy that must be given greatest consideration by any organization include:

- *Taxation*:   Taxation affects all profit-making organizations, although the levels vary from country to country. The annual net profit is the amount at the disposal of the organization after taxes have been paid.
- *Interest rates on loans*:   Most organizations depend on borrowed money to keep their business going, e.g. they use borrowed funds to purchase materials and pay for services until they receive payment for work done. When interest rates rise, the higher cost of borrowing must be covered by increased sales or they will cause a reduction in profits.
- *Level of grants*:   For various reasons, many activities depend on some support from the government. Typical examples of activities needing government support include problems of public interest requiring solutions and the attraction of new business from overseas.
- *Regulations*:   Government regulations can considerably affect the performance of an organization.

The impact of taxation

For business activities, by far the most significant government decisions are those concerned with taxation. Most individuals are aware of personal income tax, but business activity is subject to other forms of taxation. Most governments impose a tax specifically directed at business activities, such as the UK government's corporation tax, and there may be additional levies on the profits of specific activities, such as petroleum production. In addition, every registered commercial concern in the European Community has to make provision for processing value added tax (VAT). Similar forms of indirect taxation are to be found in most other countries as well.

The impact of government regulations

The government legislation with the greatest impact on engineering is in relation to building or manufacturing standards and safety regulations. This is particularly true of the latter. The steady demand for an increased level of safety in, for example, passenger cars, has a double-edged effect on the manufacturers. The provision of innovative features such as a collapsible steering column will usually increase sales in the long run as well as contributing to the well-being of drivers, but in the short term it can lead to cost increases and a temporary drop in sales.

Other problems for which government regulations are forcing producers to find innovative solutions include, for example, those of air pollution from car exhausts and the environmental effects of the dumping of waste in the sea.

The effects of interest rates

Most organizations have to borrow money from time to time to fund the initial stages of an operation, to facilitate expansion or open up new ventures. Such borrowing often has to be paid for by means of interest on the loans obtained.

The first point to note is that the interest rate on borrowed money is usually higher than on invested sums and over a period of time there will be fluctuations. Usually the rates drop as more money becomes available, and rise as such funds diminish. Hence it is during periods of low interest rates that major expansion projects and fresh ventures are likely to be undertaken, while organizations are naturally less likely to take on new commitments when the cost of borrowing is high.

A second point to bear in mind is that the government of a country may choose to manipulate the interest rates operating in its own money market in order to control the state of the economy. Such changes can have an impact not only on the operations of commercial concerns within the country but on export companies in other countries trading with it.

The impact of government expenditure

In many countries government expenditure offers commercial opportunities to various sectors of industry. Typical examples of areas in which government spending has had a major industrial impact include the following:

● *Health*: The building of new hospitals and the provision of advanced equipment and training.

- *Transport*:   The building of new roads and the maintenance of existing ones.
- *Education*:   The construction and equipment of schools, colleges and universities.
- *Defence*:   The placing of orders for new aircraft, weapons and warships.
- *New ventures*:   The exploration of space, which requires new materials and new technologies.

All of the above situations would provide a considerable number of opportunities for commercial organizations.

Changes in government policy

Circumstances that may lead to a significant change in the policy of a country's government include the following:

- A change of government through a general election or the appointment of a new party leader.
- The ousting of the ruling party by means of a coup or during a period of armed conflict: in either case there may be a sudden major change of policy or indeed a total reversal of the previous mode of government.
- The approach of a general election: at such times policies which are clearly not achieving the support of the electorate may be radically adjusted in order to ensure a majority vote.
- Pressure from a supranational organization, such as the European Community, can affect the agricultural and environmental policies of individual member countries.

Generally speaking, once a particular aspect of policy has been established by statute it is difficult to alter because the change involves a legal procedure. The implications of proposed new policies need to be carefully considered, therefore, to ensure that anomalies are detected and dealt with before they are enshrined in laws.

ILLUSTRATIVE EXAMPLES

Example   **5.6(a)**

Subject   **The effects of taxation**

Background   Government decisions on taxation can have a major effect on the attractiveness or otherwise of individual business propositions. This effect is illustrated here by consideration of three types of taxation operating in different countries.

Petroleum revenue tax: From the early 1970s up to the present the UK government has been deriving a great deal of revenue through taxing the profits of the oil and gas industries in the British sector of the North Sea.

Two main taxes are imposed on the companies involved. The first is corporation tax, which is applicable to all companies operating in the UK, and the other is a special petroleum revenue tax (PRT). The technicalities of PRT are quite detailed, but the two main factors are that it is applied to profits after expenses are deducted, and that it can be varied from year to year, depending on the current price of oil. The following table shows variations over a 12-year period:

Year	Rate	Average crude-oil prices (US$ barrel)
1975	45%	12.60
1979	60%	15.45
1980	70%	29.75
1983	75%	33.50
1984	80%	29.90
1987	75%	15.00

Clearly, when oil prices started to drop with a corresponding drop in profits a high PRT would inhibit oil companies from prospecting for and exploiting new sources of offshore oil. Such activity is necessary in the long term, however, and the government can adjust the PRT accordingly when necessary. This was the case after the 1986 oil crisis when prices eventually fell from US$35 to as low as US$10 per barrel.

'Withholding' tax on foreign companies: During the 1980s many European and other companies acquired US subsidiaries as a speedy means of increasing their growth. Typical European companies making such US acquisitions included Unilever, Hanson, Hoechst. Recently, however, the US Congress has put through legislation to impose a 30 per cent 'Withholding' tax on any foreign company selling a US subsidiary, i.e. a tax of 30 per cent on the profit from the sale. Parallel with this is a plan to stop tax-deductible interest payments from US subsidiaries to foreign companies. The combined effect of these measures will be that foreign companies will find investment in US business concerns less attractive.

Value added tax (VAT):   Many governments impose a value added tax on goods sold and services rendered. In the UK, for example, up to 1 April 1991 VAT was charged at a rate of 15 per cent of the value of the goods or services concerned. It was subsequently raised to $17\frac{1}{2}$ per cent. This brings in a large amount of revenue but it is unpopular with both parties in sales transactions. Customers dislike the extra cost and providers have to cope with the extra bookkeeping involved.

Comments The level of taxation in a particular country at any time can affect the profits of a company operating there and hence its response to a given market situation. It may take only a very small rise or fall to determine whether a business venture will go forward or not.

Source Press comments.

Example **5.6(b)**

Subject **An initiative by a national government**

Background Government initiatives can have a significant impact on the performance of business activities, and in particular that of small business enterprises. An example of this is the 'Enterprise Initiative' taken by the UK Department of Trade and Industry (DTI) in January 1988.

The background to this is the planned establishment of the 'Single European Market' in 1992. By the end of that year within the European Community the twelve member states (Belgium, Denmark, France, Germany, Greece, Ireland, Italy, Luxembourg, The Netherlands, Portugal, Spain and the UK) will have dropped all the barriers that affect trade between nations such as national regulations and technical standards. This will create a vast open trading region in Western Europe. Companies in every country will find that their market for goods and services has increased, but so has the competition. To encourage UK companies to become more competitive, the DTI's Enterprise Initiatives offer assistance in the following areas:

- Consultancy
- Marketing
- Business planning
- Finance and information
- Quality
- Design
- Manufacturing
- Regional
- Export
- Research and technology
- Enterprise and education

For most of these studies half the cost of consultancy will be covered by the DTI up to a maximum of fifteen persondays, and in certain regions as much as two-thirds of the cost may be underwritten. For example, under the Business Planning Initiative, 3i Enterprise Support Ltd will assist a company in formulating appropriate business goals. It will also provide guidance on the devising of business strategies and better management of design, quality, manufacturing, purchasing and supply.

Comments   Governments can provide encouragement and support for business activities in a number of ways. If a company is to benefit from a particular scheme the management must study the regulations carefully to ensure that the proposal they submit obeys the rules laid down for that scheme.

Source   *Enterprise Initiative*, UK Department of Trade and Industry, HMSO, January 1989.

Example   **5.6(c)**

Subject   **Monitoring of company mergers**

Background   There are many ways for a business to expand its activities, but one well-established approach is by means of mergers or takeovers. In theory, bringing together organizations with complementary expertise, facilities and activities will increase the effective overall performance of the whole group. In practice, a merger often involves one company taking another over for its own benefit. This may be to eliminate competition, to acquire missing expertise/facilities, to diversify its interests or in some other way to implement its objectives.

The background to many mergers is very complicated, and they may have a potentially adverse effect on national interests such as employment and sensitive defence capabilities. Many governments have therefore set up special bodies to oversee these activities, and some typical examples will now be considered.

United Kingdom: The Monopolies and Mergers Commission regulates takeovers in the UK. It can go into action for any one of several reasons. In a typical situation a company not wishing to be taken over by another will request that the bid be referred to the Commission on the grounds that a 'monopoly' situation is likely to occur in a particular industrial sector. The Commission can respond in various ways. It can investigate the complaint, refuse to allow the bid to go forward, or impose certain conditions on it. In a recent case, for instance, the General Electric Company (UK) and Siemens (Germany) had to overcome objections relating to defence capabilities before their joint bid for Plessey (UK) could go ahead.

European Community: The European Community consists of twelve countries, each with its own regulations on competition. Provision has, however, been made by the Commission to investigate situations in which it is believed that these regulations have been breached, e.g. if a company seems to have gained a monopoly position within the Community. Circumstances often make it difficult for the Commission's officials to act before a merger has actually taken place, but there are regulations to deal with particular situations. For example, if a projected trans-national takeover involved a combined turnover of five billion or more European Currency Units (ECUs) the plans for the merger would have to be referred to the Commission.

The United States: Business mergers and takeovers are constantly taking place in the USA and often on a much larger scale that in other countries. There are, however, many anti-trust regulations which prevent companies holding a monopoly position. Thus, if a company, knowingly or otherwise, increases its value beyond a given figure it is obliged to disinvest, or else it could face serious charges leading to major penalties for breaking the anti-trust laws.

Comments   Many business mergers in the past have had a crucial and wide-ranging effect on national and international affairs. Regulations imposed as a result of this should have a significant effect on how organizations now operate.

Sources   Press comments.

5.7 MATERIAL FOR FURTHER STUDY

Barrow, C. and Barrow, P., *The business plan workbook*, Kogan Page, London, 1988.

Brigham, E.F., *Fundamentals of financial management*, 3rd edn, Dryden, New York, 1990.

Brownrigg, M., *Understanding economy*, Addison-Wesley, Wokingham, 1990.

Chadwick, L., *The essence of management accounting*, Prentice-Hall, Englewood Cliffs, NJ, 1991.

Dyson, J.R., *Accounting for non-accounting students*, 2nd edn, Pitman, London, 1991.

Fleming, I. and McKinstry, S., *Accounting for business management*, Harper Collins, London, 1991.

Ibbetson, P., *Raising business finance*, Northcote House, Plymouth, 1987.

Sherret, P.N., *The complete guide to fund raising*, Mercury, London, 1988.

Sleight, S., *Sponsorship*, McGraw-Hill, Maidenhead, 1989.

West, A., *A business plan*, Pitman, London, 1988.

CHAPTER 6

MANPOWER

Six topics have been selected for consideration in this chapter. It begins with a section entitled 'Human resources' which considers the crucial place of human skills and experience in business activities. This is followed by a section touching on various aspects of 'Communication skills', from the preparation of brief reports to the effective use of visual aids.

The next two sections deal with the important complementary topics of 'Team work' and 'Leadership'. All engineers need to understand and acquire skill in working as team members and most expect eventually to have some form of leadership responsibility. Issues examined range from team composition and the efficient use of meetings to the winning of respect and factors in the motivation of personnel.

The contribution of education in the development of an engineer is considered next and the importance is emphasized of developing competence, confidence and communication skills simultaneously. this is followed by an examination of 'Time organization' and planning for the effective use of time.

The chapter ends with a list of material for further study.

6.1 HUMAN RESOURCES

Goal

To understand how best to use valuable human resources in business activity.

About the subject

The human resources of any organization are, without doubt, one of its most valuable assets. Even in the most highly computerized enterprise, human beings are supreme, as they are needed to identify markets, control the manufacture of products, handle finance and manage every aspect of the organization's activities. The quality of its human resources is therefore very important, and the same judgement and care should go into their deployment as goes into that of any other investment. Those in leadership positions need to know the abilities, qualities and personal attributes required in the different sectors of their organization's work, and be able to recognize the presence or absence of these in existing and potential employees.

Critical requirements within an organization's human resources include the following:

- Professional competence and expertise
- Communication skills
- Confidence
- Leadership quality
- The capacity for teamwork
- Experience and judgemental skill

It can be assumed that very few people possess and are able to use all these requirements to the full. There is therefore a need to utilize the organization's human resources in such a way that the best results are obtained through the linking of complementary expertise, skills and personalities.

Generally, the human resources of an organization are arranged in the form of a pyramid, as shown in Fig. 6.1. Recruits entering at the bottom of the pyramid are regarded as a part of a 'reservoir', and should, eventually, move upwards in due time. There is, however, less room at the top and only a limited number of people reach senior positions. Within the pyramid there is one layer, sometimes called the 'power house', where most of the key information and decision options are generated for action by the top management. It is also within this layer that the major source of influence lies and the performance of the organization is determined.

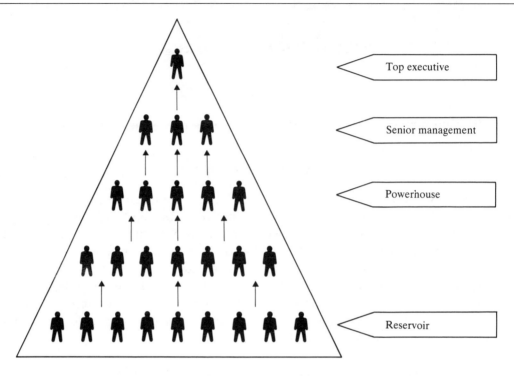

Figure 6.1   An organization's human resources forming a pyramid

Key questions and issues

The financial value of human resources

Companies are required by law to compile and publish their accounts at regular intervals, usually twelve months, but very few ever attempt an 'audit' to assess the value of their human resources. This seems surprising because, while equipment and facilities lose value with time—or 'depreciate'—through wear and tear or obsolescence, people generally become more valuable to the organization as they gain experience. Indeed human resources are normally the only 'asset' that appreciates with time! This is why there is a strong case for assigning a financial value to human resources or defining their 'equivalent asset value'.

The qualities desirable in an engineer

'Can you provide us with a good engineering graduate?' This is not an unusual question from an industrialist to friends in the academic world. The enquirer may not define what is meant by 'good' but the word usually implies a graduate who has a wide range of knowledge. This, however, is not sufficient. A high quality engineer is someone who

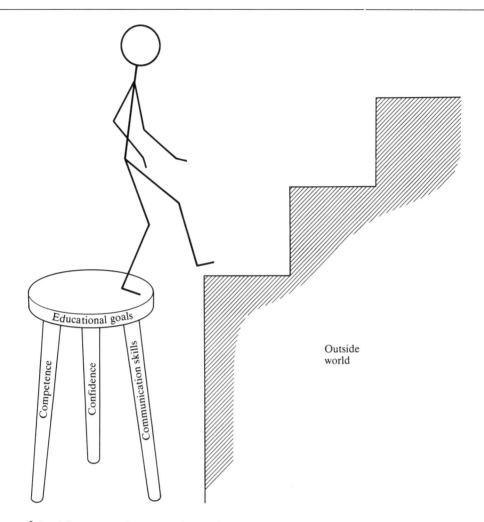

Figure 6.2   The 3-C educational stool

possesses a balance of competence, confidence and communication skill, as illustrated in Fig. 6.2. More specifically, these attributes involve the following factors:

Competence: This involves the ability to:

- understand fundamental principles
- acquire, access and apply knowledge
- generate and develop ideas
- identify and prioritize options
- solve problems and make decisions

Confidence: This involves the ability to:

- select realistic goals
- organize time effectively
- develop positive attitudes
- be flexible
- take on responsibility

Communication skills: This involves the ability to:

- transmit and receive oral, written visual and physical communications
- overcome barriers, leakage and interference in the communication medium

Engineers as managers

Some hold the view that his or her bias towards technological subjects prevents an engineer from becoming a good manager. The same commentators would even go so far as to say that, because an engineer's training has focused so much on specialist topics, an engineer–manager often cannot see the total picture of an organization. This in turn leads to a tendency to promote the aspects related to his or her own interests rather than implementing the corporate objectives of the organization. There is an element of truth in this impression, since the majority of engineers do not expect to become managers until a fairly late stage of their careers and tend to base policy decisions on technical rather than commercial factors.

Arguments against the effectiveness of engineers in the management role, however, overlook the fact that we live in an age of increasingly complex technologies. Sound commercial decisions about engineering activities cannot be made without some understanding of technological principles and practice. It is also useful to have the ability to perceive the possible serious pitfalls inherent in a 'well-packaged' proposal.

A positive attitude to change

The staff of an organization can only continue to contribute effectively if their abilities are properly used as they gain experience, and their skills are being regularly upgraded. Even more important, however, is provision to help them adapt to change. A positive attitude needs to be developed so that new practices are flexibly embraced, as they are usually introduced to keep the organization abreast of technological advances. Resistance to change may arise from various causes:

- Fear of the unknown (often stemming from a sheer lack of information)
- Unhappy previous experiences of redundancy or failure to cope
- Lack of forward planning for the change(s)
- A combination of the above factors

If even a minor change in work practice is imminent, a concerted effort must be made to help all concerned to take a positive attitude towards it. This may entail making available information such as market research results so that the proposed change can be seen in context and its implications appreciated at an early stage. It will also call for the exercise of good leadership, with managers taking time to explain the proposed changes in detail to those directly affected and negotiating positively for their implementation.

Characteristics of an effective manager

An effective manager is one who maintains the right balance between the two characteristics of being enterprising and being effective in meeting goals. Enterprise is necessary to initiate the new ideas and fresh approaches which are crucial to success in today's competitive world. Too many such initiatives, however, can be destructive, because resources are overstretched, resulting in failure to reach one objective before heavy commitment to new schemes. On the other hand, well-disciplined organizers do get work done, as they are less likely to be sidetracked by new ideas before the current goal is achieved. They are, however, usually less able to generate new initiatives. It is important, therefore, to help people develop both characteristics, or else to create a good mix of those who are strong in one or the other.

ILLUSTRATIVE EXAMPLES

Example   **6.1(a)**

Subject   **One Company's management of human resources**

Background   IBM (International Business Machines) of the USA is the world's leading computer company, which has dominated the information technology and data processing market since the mid-1960s. The company's policy is to employ high-quality staff, including a significant number of engineers, and its level of staff care compares very favourably with that of other similar organizations. IBM has always recognized the value of human resources and the following aspects of its approach are worth considering:

- *Full employment policy*: Between 1965 and 1990 no employee has been paid off through lack of work. During periods of recession, surplus staff in unprofitable divisions have been transferred to others more profitable. During periods of short-term expansion, the policy has been to meet the surge in demand by means of overtime, subcontracting and temporary workers. This approach has led to a strong sense of company loyalty among the workers.
- *Commitment to training*: IBM invests heavily in training and personal development programmes because it recognizes that, without continuous development, human resources—like any other asset—are likely to become out of date. Accordingly, a typical annual training budget during the mid-1980s was in excess of $500M. Investment of this order has

enabled IBM to maintain a very sharp edge in an extremely competitive industry.

- *Deployment of expertise of highly-qualified staff*:   At the start of the 1990s, IBM recognized a problem which stemmed indirectly from the successful result of its employment policy, i.e. a low turnover of staff. Combined with rapid expansion in the 1960s this low turnover meant that the company had a large group of highly qualified executives who were over 50 years old but who were reluctant to retire early and thus curtail their active working life. The IBM solution was to form a consultancy company called Skillbase in July 1990. This venture allows employees taking early retirement the opportunity of a new career in consultancy with the minimum of financial risk. IBM offers them an attractive package which includes: a year's salary, no reduction in pension value at 65, and a guaranteed ninety days work per year for two years at 40 per cent of their previous basic pay. But this means it can still call on the valuable experience and expertise of these people while giving them freedom to undertake additional work with other organizations via Skillbase.

Comments   IBM's policy is based on an understanding of the fact that an organization's performance is dependent, not simply on knowledge of its markets, effective management and adequate financial resources, but also on a team of loyal and high-quality staff.

Source   *Employment with IBM: Principles at work*. Booklet published by IBM. Golzen, G., Trimming down payrolls the IBM way, *Sunday Times*, 26 October 1990.

Example   **6.1(b)**

Subject   **Retaining human resources**

Background   'Head-hunting' is the term used to describe the activities of organizations which specialize in finding key personnel for their clients. The importance many organizations attach to this activity is illustrated by the amount they are prepared to pay for this service. The costs include search fees, senior executives conducting selection interviews, relocation, introductory training on company working practices, as well as other expenses. This total could be equivalent to the first year's salary of the recruited person, and there is no quarantee that he or she would be the right person for the job.

More recently, the term 'head-fastening' has been used to describe the retention of key staff, i.e. the reverse of head-hunting. Many companies now realize that it is more effective to retain experienced staff and have introduced incentives and contracts to persuade such staff to remain with the organization.

A typical example of a very attractive package offered by one multinational company in order to retain engineers with more than eight to ten years of working experience includes the following:

- *Special incentive salaries*: As well as paying salaries which are above average for the industry the company has an incentive scale, ranging from the amount earned by someone newly upgraded or recruited, to enhanced rates for employees showing outstanding qualities and achievements in performance.
- *Fringe benefits*: These include a company car, a non-contributory pension scheme, good pensions, health insurance and interest-free loans over a given period for house purchase and other help when staff are asked to change location.
- *Staff privileges*: Company shares are available to staff at special rates, or else on a scheme whereby an additional share will be offered for every share bought, up to a given percentage of the purchaser's salary.

Comments   These instances are clear signs of how much companies value their human resources. It is often far better to develop existing staff than to bring in new personnel, especially when a strong sense of loyalty to the organization exists.

Source   How to hold the people you want to keep, *Management today*, November 1985.
Press comments.

Example   **6.1(c)**

Subject   **Identifying hidden skills**

Background   An international publisher located in England but with customers throughout Europe has its Commercial Department divided into several sections, such as Accounts and Invoicing. For a period some years ago the Personnel Department was aware of disharmony in the Cash Section, centring on one employee. She was very competent but, born and brought up in Eastern Europe, had a rather 'direct' way of dealing with people and this tended to upset all who came into contact with her in the section. While seeking to solve this problem the management realized that the Invoicing Section was having problems regarding prompt payment of bills by North European customers. Eventually the manager took the 'bold' step of promoting this employee from the Cash Section to the Invoicing Section with special responsibility for Eastern and North European accounts. The result was that bills were paid, on average, four weeks sooner than previously. In financial terms this represented an extra £1 million of liquid cash in any four-week period. In consequence of this success the company began examining ways of identifying the hidden capabilities of all employees in order to increase their contribution to the company by filling key posts from within the organization.

Similar experiences by other companies have persuaded them to take greater notice of the background, hobbies and personal interests of the workforce. It has been found, for example, that people doing very routine work may well have a sound knowledge of certain foreign countries, or be highly skilled at computer programming, or sailing, or architectural design, or the drawing of cartoons.

Comments   In many instances the 'paper' qualification held by an individual may not be the one of greatest value to his or her employer. It is surprising how much hidden talent within an organization can be identified when a flexible approach to staff deployment is adopted.

Source   Press comments and private communications.

6.2 COMMUNICATION SKILLS

Goal

To achieve a high level of interaction between relevant individuals in order to meet the business ojective in the most effective way.

About the subject

'Communication' is the term used to cover a number of activities, ranging from the exchange of information and the spreading of news to the transmission of data and, in the plural, for transportation links between centres of human habitation. In the present context the term is used for interaction between two or more human beings in order to exchange information.

To achieve effectivenesss in this activity communication abilities and skills have first to be identified and then acquired, and the latter is done in the same way as any other subject is learned. Communication skills are important because individuals need to interact with others at all levels, both within the organization and with the world at large.

There are four principal components in a communication 'circuit': the message, two senders/receivers and the medium for transmitting the message between them. In the situation we are considering each sender/receiver is a human being and the message may be transmitted by any one of the following media:

- Speech
- Writing
- Gestures
- Body language
- Physical contact

Using continuous dots to represent the message, the top diagram in Fig. 6.3 shows the 'ideal' communication circuit, where the message is clearly transmitted and received without any difficulty. However, in practice this is seldom the case and this figure also demonstrates a number of possible 'faults' or 'noise effects' which can inhibit effective communication. The major ones are:

- *Leakage*:   Only a part of the message reaches the receiver. For example, part of an instruction is omitted by the sender or it is lost during transmission.
- *Blockage*:   The message cannot be understood because of physical or mental blockages. A bad telephone line, for example, would cause a physical blockage to oral communication. Mental blockage can be caused by an attitude of mind or a lack of vocabulary, either of which can prevent a message being understood.
- *Interference*:   External factors can also affect the transmission of the message. A discussion can be physically inhibited by the level of noise surrounding it. An inability to concentrate on the message being communicated would constitute mental interference.

Successful communication depends on the ability to identify the objectives of the message, to quantify the needs of both sender and receiver, and to overcome any potential difficulties. In addition it is important to select the most suitable medium for transmission and to use appropriate criteria for evaluating the outcome.

Key questions and issues

Principal forms of communication

Many forms of communication are used in business but the principal ones are:

- Oral
- Written
- Visual

Each can be used either individually or in combination with the others, but the skills

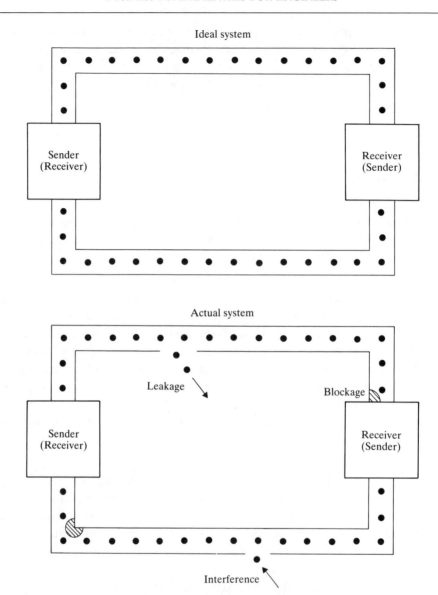

Figure 6.3   Communication circuits: the ideal and the reality

related to each need to be properly developed if effective communication is to take place. It is also useful to understand the basis of each form.

Oral communication

A great deal of business activity takes place using this form and typical examples include:

- Telephone conversations
- Formal meetings
- Presentations
- Entertaining visitors and clients
- Interviewing people for jobs and for appraisal purposes
- Negotiating contracts
- Making speeches
- Persuasive and counselling activities

Successful communication by any of these means depends on prior preparation wherever possible, knowledge of the other parties, a logical presentation of data, the most appropriate mode of communication, practice and self-confidence. One's own technique can be greatly assisted by critical observation of the 'performance' of good oral communicators in order to identify the factors that contribute to their success.

Written communication

Written forms of communication are used to ensure that there is a record of significant interaction, agreement, results, findings, reminders etc. Important forms of written activity in business include:

- Letters
- Faxes and telexes
- Reports
- Memoranda
- Contracts
- Notices

As with oral communication, success again depends on thorough preparation, a knowledge of the receiving audience, selection of the most appropriate mode, practice and self-confidence. Being able to write clearly and concisely is a vital asset here. Indeed, many 'busy' managers do not like to read more than one page and expect all the relevant issues to be covered within this space! Many people find it helpful to take classes in report writing, or to study books on the subject in order to increase their expertise.

Visual communication

Visual communication is employed extensively in business activities. This is so because a good picture can improve comprehension and also because time for the actual communication is almost always limited. Thus, a well-planned presentation using visual aids can do much to reduce the time required for delivery and assimilation of the intended message.

Depending on the situation, popular forms used include:

- Charts and graphs
- Overhead viewgraphs
- Transparencies
- Films and videos
- Photographs
- Physical objects

Success with this form depends on limiting the amount of information to be conveyed in each item. For example, a good viewgraph shows the keywords of the presentation instead of containing a photocopy of a full page of text. Likewise, a few well-chosen and well-displayed transparencies can highlight technical points, but a large number will actually bore the audience.

Major barriers to effective communication

Difficulty in communication can be attributed to a variety of causes, particularly where people trained in technology are concerned.

Firstly, the responsibility for ineffective communication may lie with the initiator. It may be due to a lack of skill in presentation, inadequate information, language difficulties or excessive use of specialized terminology. There may also be a failure to recognize that good communication is as crucial as correct technological results, or the presenter's own lack of conviction. The commonest fault is inadequate preparation. Omitting to identify the type of receiver or audience is the outstanding example of this type of failure.

Alternatively, the fault may lie with the receiver. Typical problems include: lack of listening skill, failure to allot time for receiving the information, lack of interest in the subject, a closed mind, incorrect expectations, and inability to understand—or impatience with—the terminology used.

The barrier to good communication may equally be caused by the delivery medium. The selection of a good and appropriate medium is crucial. Typical problems would be: poor acoustics in the room where a meeting is held; the fax machine omitting a critical line of information in a fax message; the projector bulb blowing during an important presentation.

The effectiveness of response, however, depends to a large extent on the implementation of the other factors, because 'errors' in any one of them will lead to unexpected responses.

Improving communication

There are many ways of improving communication skills. The principal ones involve understanding and applying certain basic elements. It is true that not all of them are required for every act of communication, but it is surprising how often an experienced communicator goes diligently through the complete set without necessarily being conscious of this. They comprise the following activities:

- *Define*: The objective of the communication should always be clearly defined at an early stage.
- *Acquire*: Background information should be acquired, including data on the receiver, the mode of transmission and the amount of time available.
- *Select*: The most appropriate mode (including vocabulary and style) for the occasion should be selected, whether writing a letter, making a telephone call, or issuing a press release.
- *Prepare*: To ensure that the envisaged 'act' will be successful it is essential to prepare thoroughly. Where possible, rehearse before friendly critics prior to delivery.
- *Deliver*: Every effort should be made to ensure that the communication is delivered with confidence, conciseness and conviction. This involves the transmitter having and demonstrating a sound knowledge of the relevant facts.
- *Evaluate*: The performance should always be evaluated against such criteria as impact, content, adherence to the time allotted, voice level and clarity. Refinements should be incorporated in the light of experience gained.

The process is summarized in Table 6.1.

Table 6.1 A procedure for improving communication

STEP	ACTION	TASK
1	Define	the objective at an early stage
2	Acquire	background information on all key aspects
3	select	the most appropriate mode(s) for the occasion
4	Prepare	thoroughly, and rehearse with a friendly audience
5	Deliver	with confidence, conviction and conciseness
6	Evaluate	the performance and refine based on experience

Interactive communication

Often communication has to take place in an interactive situation where instantaneous responses are expected. One example would be a quiz session involving question and answer between a speaker and audience. Alternatively, interaction can take place via computer terminals or the telephone.

The key difference between this form of communication and others lies in the limited amount of time for considering one's response. The sooner the response is expected the greater the pressure. Some individuals perform well under this type of pressure while others tend to dry up in such situations. There is a misleading idea that the former are 'brighter' or 'cleverer' than the latter. In fact good performance under these conditions depends far more on personality and practice. Some people enjoy the challenge of the interactive situation, and practice can enable most to increase their ability to cope with it.

Effectiveness of communication

Attempts have been made to measure the effectiveness of communication, and an accepted statistic has been identified for public speaking. The following distribution has been suggested:

Elements	Effectiveness
Words or message	7%
Voice	38%
Appearance	55%

In other words, the dress of the communicator and the way the message is being communicated make much more impact than the message itself!

ILLUSTRATIVE EXAMPLES

Example   6.2(a)

Subject   **Communication via a presentation**

Background   The UK Department of Trade and Industry launched a programme for developing resources from the oceans and invited interested parties to put in bids for performing six feasibility studies at a fixed cost over a period of six months. Proposals had to be submitted by a particular date and bidders were then asked to make supporting presentations on a given day.

The aim of the presentation was defined as 'clarification' of the written proposals before a general audience, including DTI officials and other interested parties. A total of fifteen minutes was allocated to each presentation, and this also included time for some questions.

Since there were between three and five bidders for each feasibility study the assessors and audience had to listen to 24 presentations in the course of the day. Overhead viewgraphs were used by every bidder and the presentations can be grouped under three headings as follows:

- *Too little information*: Ten of the presentations gave very little information about their proposals. In these cases the speaker talked around two or three hand-drawn viewgraphs which had evidently been hurriedly prepared, and there seemed to have been no rehearsal before the actual presentation.
- *Too much information*: Twelve more provided very detailed information on their proposals, in some cases with photocopies of typed sheets being handed out. Here it was difficult to grasp all the information presented or to be sure what the study would deliver. In most of these presentations no time was left for questions.
- *Relevant information*: Only two presentations fulfilled the actual requirements. In both cases ten minutes were used to make a presentation backed by a set of simple viewgraphs giving relevant information without confusing detail. Schedules were outlined and what would be delivered by the study was clearly stated. Questions raised were given replies which were concise and to the point.

Comments   Although a presentation of this type may not be the most crucial decision factor, the bids by both organizations in the third group above were successful. Their effective presentations served several purposes. In particular, they highlighted the key points of the proposal for the decision makers, who might not have been able to read the complete text in the time available between submission and presentation dates. The decision makers were also impressed with the kind of professionalism that would be provided if the bids were successful.

Source   Private communication.

Example   **6.2(b)**

Subject   **Communicating in writing**

Background   In many fields of activity written communications need careful attention. This could be due to choice of words, or to specialists failing to realize that their expertise is not understood by everyone. To illustrate this point two examples have been selected.

- A long dispute has been in progress between the buyers of a fleet of farm vehicles and the manufacturer. The situation reached a peak when an open letter from the managing director of the buying company outlining the faults found in the vehicles ended as follows:

  'To summarize, I find it totally unacceptable that most of the vehicles manufactured by your company have been found unsafe and unsatisfactory.'

It is not surprising that the manufacturers vigorously defended the quality and reliability of their products. The resolution of this dispute took considerable time. It could have been resolved much earlier if the managing director had selected one different word. The use of 'some' instead of 'most' would have allowed the basis for a compromise to be achieved.

- An article entitled 'General economic background' addressed by an investment specialist to potential investors among the general public reads as follows:

'There are now clear signs of a slowdown in economic activity in the US and the UK but activity remains robust, especially in Continental Europe. . . . Whether or not the much-mooted soft landing is achieved remains in the balance, but investors seem to be favouring this outcome and support equity investment.'

It is difficult to understand what a 'soft landing' is, and why support for equity investment was sensible. Not surprisingly, the response to the circular was unfavourable.

Comments	In both these examples, the key assumption was that the readers would understand what was meant. It is hardly surprising that neither communication achieved its objective.
Sources	Press comments and private communications.
Example	**6.2(c)**
Subject	**The best use of viewgraph slides**
Background	Visual aids are being increasingly used to enhance presentations as a way of improving communication. One popular visual aid is the viewgraph and Fig. 6.4 gives two examples of viewgraph slides to illustrate a presentation on the three possible systems available for controlling the position of a ship relative to an offshore structure.
	Viewgraph (*a*) is extremely full of data, and is far too complicated for the audience to take in and remember the salient facts. In practice, (*b*) was selected because it contained sufficient data to increase the effectiveness of the presentation.
Comments	Many people tend to cram too much information into a single viewgraph. This is counter-productive. It is much better to limit the information on the slide to concise headings which convey the main facts or else to use a series of key words on the slides to cover different aspects of the topic.
Source	Conference viewgraph and private communication.

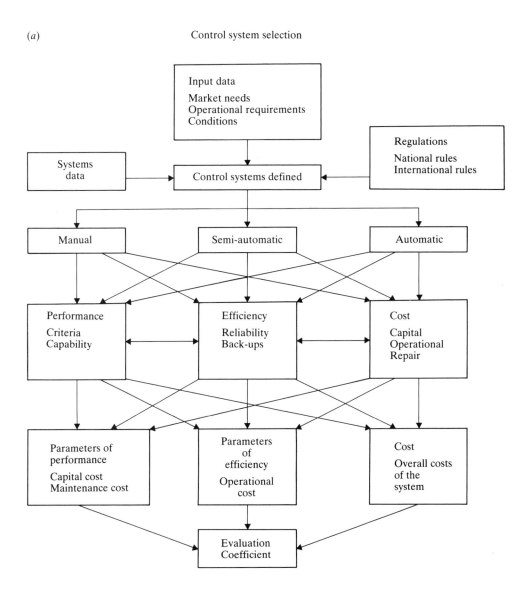

Figure 6.4 Two ways of presenting a message on (a) a bad viewgraph, and (b) a corresponding good one (see overleaf)

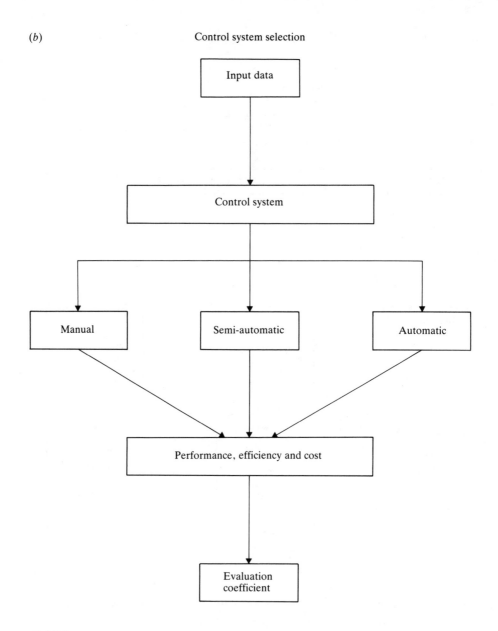

Figure 6.4(b)

6.3 WORKING IN A TEAM

Goal

To obtain better performance by means of a proper mix of expertise and personal qualities.

About the subject

'Teamwork' is a topic that is receiving considerable attention today in many organizations and activities. The reasons for this are many, and include the following:

- The ideas of teamwork and the team spirit are a good basis for motivating all those involved in an activity to contribute to their full potential.
- From a public relations angle, decisions are usually more acceptable if they can be attributed to a group of people rather than one individual.
- Teamwork should ensure a continuous flow of good ideas, with decisions being made after extensive discussion.

So what is a team? It is the term used for a group of two or more individuals brought together by some common focal point or objective. Typical responsibilities of a project team, for example, would be:

- Managing the design, construction and installation of a processing plant
- Developing regulations for improving the safety of a manufacturing process
- Organizing the welfare activities of the company

It is the need to work together to achieve the specific objective that differentiates a team from any other group of people. The success of the team's work generally depends on a sense of common identity, the establishment of good understanding between members and effective methods of communication. Its structure can be either formal or informal, but its main responsibility is to achieve the objective for which it was formed.

Key questions and issues

Forming a project team

The main reason for forming a project team in a business situation is to enable ideas and expertise to be pooled and activities coordinated so that the project will be effectively completed in the minimum of time. The composition of the team is therefore very important. The actual composition of a particular team will depend on the objective in question but it usually involves the following considerations:

- The goal to be achieved
- A spread of expertise
- Representation of different interests

A team composed of engineers from the same background and discipline would probably not suffer from communication problems. Such a team, however, would offer a very limited range of options and could be lacking in certain useful types of expertise, e.g. financial planning. On the other hand, a team where members have little common expertise would find communication difficult, as different disciplines may interpret the same words in very different ways (e.g. the mathematician's and the engineer's view of 'accuracy').

As a basic principle, a project team should include those with a mix of technological expertise and experience, and some with a financial and/or managerial background. This will enable the most effective comprehensive strategies to be devised, possible technological and financial problems to be identified and provision made to overcome them.

Team organization

There are various methods of organizing teams for specific assignments, ranging from hiring temporary staff to selecting a group from one unit within the organization.

One popular method of providing staff for projects and assignments is to build up teams of personnel with complementary skills who are based in various divisions or sections of an organization. The skills required may already be available within the organization because they are called for on a regular basis, but they are not being utilized 100 per cent of the time and there is therefore 'spare capacity' to draw on. Alternatively, a pool of special skills may be deliberately assembled by the organization to serve various needs as they arise. In recent times, the latter method has been called 'matrix management', because it is possible to tabulate projects against skills on the rows and columns of a matrix (see Fig. 6.5).

In this figure, each row represents a project while each column is assigned to a given skill. Some of these skills are required for every project while others are used more selectively, e.g. estimating skill as opposed to stress analysis expertise.

The main advantages of this approach are:

- The expertise of the different divisions and sections of an organization is employed more efficiently to meet a range of needs.
- Synergy can be expected from the team effort.
- Improvements in communication between people from different divisions and sections of the organization.

		Main sections on core skills		
		Column 1	Column 2	⋯	Column C
Projects, assignments or tasks	Row 1			
	Row 2			
	⋯			
	⋯			
	Row R			

Figure 6.5 Matrix management: tasks versus skills

The main disadvantages include:

- Practical difficulties caused by highly specialized staff being unable to match the specific requirements of a project.
- Possible duplication of expertise within an organization.

Team effectiveness

Creating a balanced team will not necessarily guarantee success, as a number of factors can influence its effectiveness. Typical of these would be:

- *Personality*: It is always wise to consider how different types of personality will work together. Conflict could arise between two individuals with lots of entrepreneurial flair. Teaming a 'born entrepreneur' with someone who has the ability to take an idea systematically through from its conception to the required end-point, however, would result in a proper mix of ability.
- *Terms of reference*: The terms of reference of the team should be clearly stated so that members and others involved know precisely what the objective is, how this is to be achieved, and what the key constraints are.

- *Team size*:   There is no 'ideal' size of team for engineering projects. The optimum for an individual project would be 'large enough for a good flow of ideas and spread of responsibilities but small enough to ensure effective communication is maintained'.
- *Backgrounds*:   Teams for engineering projects often comprise members of different nationalities, religions and customs, and these are additional factors that need to be taken into consideration, if smooth cooperation and coordination are to result.

The critical stage in working together

The most critical phase for a team is the first stage, for example, the first meeting of those who have been assigned to the team or invited to help form it. In some cases various members will have worked together before, but in others they will all be meeting one another for the first time. The latter is particularly likely in a large organization when a team comprises representatives from several different sectors. It is important at this stage that members get to know each other and freely and openly discuss the objective of the assignment, the terms of reference and modes of working. It is also at this stage that they begin to recognize each other's views, approaches and experience. A team usually has a leader, either appointed beforehand or elected by the members after their first meeting. Members must be adaptable enough to accept the role of either leader or team member as required.

The maturing process

If a team is to work effectively, members must appreciate that it will have to go through a maturing process. This will probably include some of the following typical phenomena:

- Prolonged debate on the terms of reference
- One member attempting to impose his or her will on the others
- Indecisive leadership, resulting in poor progress
- Personality clashes, leading to conflict between members
- Difficulties with language
- Unfamiliarity with the work styles of other members

To overcome these problems each member must gain the respect of his or her colleagues. The team as a whole needs to cultivate mutual trust and understanding, but a clear definition of individual assignments will do much to avoid or overcome conflict.

Measuring progress

In order to operate effectively as a cohesive unit, the team must have a method of measuring its progress or performance so that the standards set are met and group loyalty is promoted. The exact criteria will, of course, be determined by the type of assignment, but would always include the following questions:

- Has the target for each stage of the project been reached?
- How closely has the time schedule been adhered to?
- If applicable, is actual expenditure keeping in step with the planned projection?
- Have progress reports been produced on time?

Group meetings

Team meetings constitute one of the most common aspects of team work, when members come together to exchange views and report on progress. They may be held for briefing purposes and information exchange, for seeking a consensus view on an issue, for solving problems, for discussing progress, or for making decisions. It is helpful to fix a regular time for team meetings, e.g. every Friday morning, so that significant issues can be regularly reviewed. This also allows team members to prepare themselves to contribute and participate fully in the meeting.

Effective meetings

The effectiveness and efficiency of a meeting depends on both chairman and team members, and both parties can contribute to its success by adherence to certain basics:

- *Agenda*: It is always helpful to provide an agenda in advance so everyone concerned knows the purpose of the meeting and the various items to be covered. This enables all concerned to prepare properly. An agenda can be very brief, showing only the main headings, but it may be expanded to include some explanation of each item.
- *Regular scheduling*: This allows scope for preparing reports and provides a recognized opportunity for dealing with problems. It is helpful to plan for the best possible use of each meeting.
- *Length*: It is never easy to keep a meeting concise, but adequate preparation, by members as well as leaders, will help. Allocating a specific fraction of the total time-span to each item of the agenda can be a useful device for promoting brisk discussion of business.
- *Flexibility*: The chairman must ensure that there is adequate opportunity for discussing new issues and considering recent problems. Judgement needs to be developed in order to strike the right balance between being too superficial and labouring over detail.
- *Minutes*: Someone should be assigned the task of noting key decisions and future actions. Such notes need not be too detailed, provided the essential points, the decisions reached and actions to be taken are accurately recorded.

Successful meetings are those conducted in such a way as to prevent the conscious or unconscious introduction of too many digressions, and those where provision is made for following up good ideas.

ILLUSTRATIVE EXAMPLES

Example	**6.3(a)**
Subject	**Team work on a major project**
Background	Oil was discovered in 1974 by an oil company in the UK area of the North Sea. However, the quantity found, or its reserve, was not large enough for an immediate go-ahead to be given for developing the field. The discovery was therefore put on the shelf. Over the next sixteen years, the file was periodically taken down and feasibility studies carried out on the reservoir without a firm conclusion being reached. Towards the late 1980s, prospects began to improve due to a combination of factors: the management's view of the smaller reserves within a varied portfolio, advances in technology, large reserves not being discovered and the realization that development costs had to be reduced.

Thus early in 1990, a project team was given the task of producing a plan for the development of this field. It should be appreciated that in oil development projects the team can comprise between 10 to 1000 members during the various stages. In the present context the team consisted of the company's own staff who were involved in the first year of activities and involved 15 to 30 members. The project team was organized in a way which differed from the company's normal approach and the key features are highlighted under the following headings.

Formation of the team: The team members were intially selected based on disciplines required to develop the field. The total was 15 with three 'mainstream' engineers plus specialists in areas such as geology, reservoir, taxation, processing, finance and environment. The individuals were nominated by the appropriate disciplines and the criteria employed included: suitability of candidates and their availability; the importance the sections attached to the project; and the stage of the candidates' careers. The team also selected its own leader.

Contributions of the first meeting: The first meeting made two vital contributions. Firstly, it reached agreement on the objective and it was to demonstrate by the end of 1990 that the field could be economically developed. Secondly, the members decided on how the team should be organized. The members unanimously rejected a pyramid-type hierarchical structure and, instead, agreed to a Venn diagram type of arrangement (see Fig. 6.6). The three skill areas were commercial, facilities and subsurface.

The outer ring consists of external organizations, such as project partners, government agencies, regulatory bodies, suppliers and subcontractors. Throughout the period in question, the objective and basic operating model were maintained.

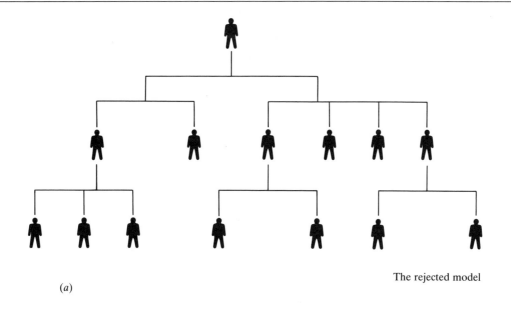

(a)

The rejected model

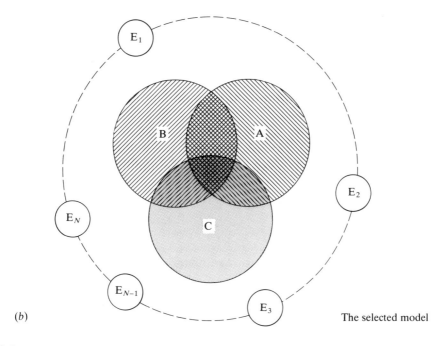

(b)

The selected model

Figure 6.6 Two possible models for organizing a team. (a) shows a hierarchical model, while (b) shows a more flexible arrangement, E_1, E_2 ... are external organizations and A, B, C are the core skills.

Methods of working: The actual methods of working were similar to those popularly used in teamwork, such as meetings and the circulation of notes. The additional component was the realization that each member had to adopt a different 'mode of behaviour'. It was necessary to iterate many times and each time reinforce the common objective and seek out fresh behavioural possibilities. Typical examples include: understanding the role each member plays in contributing to the project and to each other; speaking openly, especially on issues which offered potential conflicts; and being positive at the meetings and seeking alignment around the common objective.

Maturing: Teamwork effectiveness increased as experience was gained and fresh attitudes developed. The maturing process was best illustrated by the way in which the exchange of information was achieved. At the beginning, everyone wanted copies of all the available information. In time, members sought and provided only the relevant information for undertaking specific tasks. Indeed, the team devised a principle on information exchange which can be represented by two questions: 'What do I need?' and 'What do they want?'.

It is therefore not surprising that the project has progressed well with a strong team spirit being generated.

Comments	This project provided a challenge to the oil company which had strong traditions and an opportunity for a fresh approach to teamwork. The main lesson learnt can be stated in the following way: teams will not realize their full potential for improved productivity by simply introducing new organizational structures, they also need to develop new modes of behaviour. It is equally important to devote effort to encouraging fresh attitudes by everyone within the organization.
Source	Private communication.
Example	**6.3(b)**
Subject	**Building up teams by the matrix management approach**
Background	An engineering consultancy company is used here to illustrate the matching of expertise with specific requirements by means of the matrix management approach. The company has five main sections, as follows:

- *Engineering*: Activities include design, analysis, experiments, project management and concept development
- *Experimental facilities*: Responsible for all the actual experimental work

- *Marketing*: Mainly involved in sales activities and the monitoring of tenders
- *Accounting*: Responsible for cost control and estimating for projects
- *Administrative support*: Deals with all the administrative requirements of the company

The company's business interests lie in seven areas of activity, as follows:

- Feasibility studies in the area of dynamics
- Stress analysis using advanced computer software
- Stress measurements in laboratory and full-scale
- Vibration calculation and measurement
- Equipment testing for vibration
- Advanced concept development for clients
- Project management for clients.

Table 6.2 shows which sections provide the necessary expertise for various different types of project.

Table 6.2 The expertise contributed by different sections of a company to an engineering project

Type of project assignment	Sections				
	Engineering	Experimental facilities	Marketing and sales	Accounting	Administrative support
Feasibility studies	●		●	●	●
Stress analysis	●				●
Stress measurement	●	●			
Vibration	●				●
Equipment testing		●			●
Advanced concept development	●		●		
Project management	●		●	●	●

Comments	In this case the key section is Engineering, while the others provide active support. Teamwork in the various projects is most likely to be effective if those from the Engineering Section can give a positive lead.
Source	Private communication.
Example	**6.3(c)**
Subject	**Teamwork activities**
Background	Teamwork is often used by organizations to assist in solving problems and generating ideas. Many different techniques are employed, some of which are well established, understood and regularly used. Two typical examples are given here of techniques used in engineering project work.

Brainstorming on paper: Using this approach the participants are asked to write down three to five reasons for a given problem or state of affairs. The sheets are then collected and displayed in order to identify what most people see as the major cause of the problem.

A typical topic could be: 'Give two reasons why a new product is failing to meet the planned sales target.' A selection of the suggested reasons could include:

- Poor design
- Too expensive
- Delivery delays
- New competitors
- Unexpected customer resistance
- Inadequate promotion effort
- Unrealistically high target
- Incorrect timing
- Increase in the interest rate
- Some exceptional circumstance

Oral brainstorming: This is similar to the previous approach except that the ideas generated are expressed orally and written on a large display as they are put forward. No discussion takes place until all the team members have expressed their ideas during, say, the first half-hour. A typical task could be to seek the most efficient methods of supplying an offshore installation on a site five kilometres from the coast in a hostile environment. Initial suggestions might include:

- Suspension bridge
- Series of floating pontoons
- Shuttle ships

- Submarines
- Tunnel
- Balloon
- Helicopter
- Parachute drops
- Cable car

Once the list has been compiled the team then evaluates each suggestion and puts the most appropriate forward to the management with supporting reasons.

Comment   In general, team effort will produce a wider range of sound possible solutions than any individual can offer. However, the effectiveness of teamwork 'brainstorming' is dependent on both the professional experience of the team members and their familiarity with the technique.

Source   Private communications.

6.4 THE LEADERSHIP ROLE

Goal

To ensure that the members of a group, team or organization can work well together in order to meet defined objectives.

About the subject

The subject of leadership is concerned with the roles of individuals at all levels within an organization. Its definition can be stated as follows:

> The efficient organizing, guiding and encouraging of human resources so that given objectives can be achieved.

The question of the effectiveness of leadership is highly complicated because human roles are affected by a large number of variables. These include national attitude, culture, ability, confidence, personality, commitment, ambition and opportunity.

There are countries, such as the USA, where the declared ambition of many from an early age is to be the boss of his or her own business. In other countries nobody wants to take on responsibility unless driven by the hope of material advantage or 'power' to make decisions. Generally, however, the situation is not so clear-cut and leadership capacity is needed in almost every situation, from team sports captaincy to the management of a multinational organization.

Whether it happens formally or informally, in any activity that involves more than one person somebody has to take the responsibility of leading to ensure that the objective of the activity is reached or the desired result is achieved. With more complex situations, such as a company involved in manufacture, the viability of the company in a competitive market is crucially dependent on the quality of its leadership.

Key questions and issues

Exercising leadership

This issue—sometimes referred to as 'leadership style'—has been receiving considerable attention because in a number of countries the method of leading is often much more important than the objective to be reached.

The various leadership styles can be classified under the following broad headings:

- *Autocratic*:   In order to meet a specified goal, leadership is exercised by one individual or a small group of individuals with little consultation with other people. Generally, a high profile is sustained by the leader who is personally involved in many of the decisions taken. The 'autocratic' style is usually associated with a 'lead-by-example' approach.
- *Democratic*:   In this approach, extensive discussion takes place before decisions are made, and these are often based on the ideas that have widest support. Another term for this is the 'consensus' approach.
- *Free rein*:   The leader usually maintains a 'low profile' and encourages team members to take initiatives within an agreed framework. The leader's influence here is indirect and the approach is often described as 'leading from the rear'.

There is, in fact, no single ideal way of exercising leadership, and the efficiency of a particular style will depend on the situation in which it is exercised. Many studies have been done in actual work situations: see, for example, the work of White and Lippit as discussed by Kast and Rosenzwlig.[1] This has led to the concept of selective leadership for different situations. Figure 6.7 plots the level of efficiency against the three styles for three situations. Clearly, an emergency situation requires firm decisions and the free-reign approach would be ineffective. On the other hand, highly sensitive issues are best dealt with by a democratic approach and idea generation is best achieved under a free-rein style of leadership.

1. *Organization and management*, McGraw-Hill, New York, 1985.

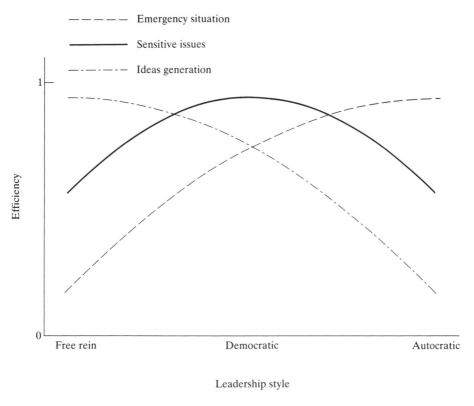

Figure 6.7 The efficiency levels of three leadership styles

The basic requirements for effective leadership

There are a number of basic factors which can help towards effective leadership in business.

Firstly, a leader should seek to acquire an understanding of the environment in which the organization is operating, and to identify any advantageous factors or constraints.

Secondly, the method of leadership selected must be one which will meet the managerial objectives, using the personal strengths of those taking the lead. It should be remembered, for example, that a North American style of leadership might not be suitable for companies operating in Asian countries. Often the most suitable style is a composite of two or more approaches.

Thirdly, a leader must be sufficiently flexible in dealing with people and may well have to modify a personal working method or style in order to suit the circumstances.

Lastly, an effective feedback mechanism must be devised which is capable of keeping the leadership fully informed and sensitive to views of the team members.

Transplanting styles of leadership

Many different styles of leadership are to be found throughout the world, but the style most prevalent in a given country normally relates closely with the nation's traditions, culture and major religion. In the 1970s and 1980s, for example, the success of Japanese industry was attributed to a culture in which detailed production planning is effectively linked with traditional attitudes to work and workforce loyalty. It was believed, however, that the Japanese style would not work effectively elsewhere. In fact, the same basic approach has been notably successful in Japanese ventures abroad, with car assembly plants in both the UK and USA achieving high profitability and industrial harmony. It would appear, therefore, that this style of leadership can be transplanted more successfully than was originally believed possible.

Various methods are employed for selecting leaders at different levels of an organization. These can range from direct appointments by means of a formal or informal procedure to the emergence of natural leaders within a group during the organization's development. Whatever the selection procedure employed, the organization's success will depend on the ability of its leadership to win the respect and cooperation of members and external contacts alike.

Winning respect

Winning respect is something that is done by different means in different situations, but there are some basic 'principles' that should help to achieve it.

Firstly, action taken by the leadership should be fair, and be seen to be fair, at all times. Apparent favour shown to one individual or one section of the organization, for example, can be misinterpreted or resented, and will undermine respect.

Secondly, decisions should be consistent, so that policies can be logically implemented. It is unhelpful for those in leadership to be continually modifying decisions or deviating from an accepted mode of operation by allowing 'special cases' to bypass the normal procedures.

Thirdly, one must ensure that sensitive issues are carefully thought through from beginning to end, with sufficient consideration given to feasible solutions for all possible outcomes. Even a major change in operating practice can be successfully introduced if it is carefully considered and discussed by all concerned before a policy decision is made.

Lastly, the difference between respect and popularity must always be borne in mind. To enjoy both would be the ideal, but this is not always possible, and a necessary but unpopular decision may not win friends but can often earn respect. For example, changes in working hours to cope with competition might not be popular but ensuring continued employment by doing so would win the respect of all.

Motivating people

Good leaders are often thought of as those who can motivate others to achieve an objective. People working for an organization may be motivated by one key factor or by a combination of several, and these may change with time and circumstances. The most obvious motivating factors are:

- Monetary reward
- Recognition of contribution
- Advancement, through being part of a successful team
- Power for decision making
- The sense of making a contribution to the organization and for society

It will be noted that money is often used as an incentive for increased 'productivity' but this must be regarded as a short-term 'motivator'. Incentive payments that are made on a regular basis will soon cease to be incentives.

Effective motivation depends on recognizing the particular stage reached by different individuals or groups of people at a particular time and using this as the basis for a set of motivating factors. For example, a company will hardly obtain the best contribution from someone who has moved to a new work location if he or she has not yet found suitable accommodation for the family or schools for the children. Assistance in over-coming these problems is often an investment leading to higher motivation in the future.

Understanding human needs

A good leader has to have an understanding of the basic human needs. The most popular theory, by Maslow, classifies these into a series of five levels beginning with the minimum.

Level 1 (Physiological):	The need for basic necessities of life such as food, clothing and living quarters.
Level 2 (Psychological):	The need for a degree of security from employ-ment, e.g. sickness and pension schemes.
Level 3 (Social):	The need for a sense of belonging, having friends and being popular.
Level 4 (Personal):	The need to fulfil personal ambitions, great and small, e.g. competence in a given skill, status in society, recognition for an achievement.
Level 5 (Communal):	The need to feel one has made a contribution to society as a whole, for example, through devising a solution for food shortages, or improving the quality of life.

The most effective decisions about the deployment of people and productivity enhancement at a particular time will be made by leaders who recognize the level reached by individuals or groups of people. It is thus possible to use this knowledge as the basis of their motivation strategy.

Additional qualities

Most engineers who become leaders of business organizations are technologically competent but they need other qualities and skills as well in order to be successful. These would include some of the following:

- Communication skills
- Administrative skills
- Awareness of the work environment
- An understanding of human psychology
- The ability to use contacts effectively
- The ability to speak well and with confidence

The ability to be in the right place at the right time can also contribute greatly to success.

ILLUSTRATIVE EXAMPLES

Example	**6.4(a)**
Subject	**The effectiveness of different leadership styles**
Background	An experiment performed by Kurt Levin on the effectiveness of leadership styles, involved three groups each led in a different way.
	Group A had an 'autocratic' leader who determined the policies, selected the work and assigned tasks to each member. Group B had a 'democratic' style of leadership with every member actively involved in running group activities. Group C had a 'free-rein' leader who took no control over the group, and members worked on their own initiative.
	The results of the experiment were that Group A, although quarrelsome, was able to work efficiently, especially when the leader was present. The members of Group B got on well and work progressed smoothly even when the leader was absent. Group C had lots of ideas but its performance efficiency was low.
	This experiment led to the concept of selective leadership for different situations.
	In practical situations the issue of leadership style is not so clear-cut. To a

large extent, the success of a project will depend on how far the style adopted is appropriate for the type of objective involved. This is best demonstrated from the record of production on the UK Continental Shelf during the period 1970 to 1990.

First phase 1970–80: The trebling of oil prices in 1973 caused what is known as the first oil crisis and the national energy policy adopted by the UK at that time was summed up as follows: 'Self-sufficiency in oil by the end of the decade.'

The successful project leaders at that time are best described as 'goal-getters'. They recognized that deadlines had to be achieved, regardless of cost, and usually opted for proven techniques and a leadership style that was more or less 'autocratic'.

Second phase 1980–88: The cost of producing oil offshore had risen greatly due to a number of factors, which included newly discovered fields having relatively low reserves; the very high cost of maintaining offshore installations; the emergence of fresh technological solutions; and a sudden drop in oil prices.

With self-sufficiency achieved, the objective of the UK's oil producers became: 'To reduce production costs and achieve cost-effectiveness.'

There was no let-up in the need to meet project targets, but in order to achieve these new objectives project leaders had to be willing to examine alternative solutions even before they were 'totally proven'. Team members became much more involved in the strategy for achieving goals and the most popular leadership style at the time was a combination of the 'democratic' and the 'autocratic', with the former predominating.

Third phase 1988 onwards: Following the major disaster on the Piper Alpha platform, with the loss of 187 lives, the emphasis has shifted to safety, and the current policy can be stated as follows: 'Safety will be of paramount importance in offshore operations.'

To achieve this goal requires a combination of education, the integration of safety requirements into every stage of a project and quality implementation of every activity.

Neither the autocratic, free-rein or democratic style of leadership seems to be appropriate here, and success would appear to require a new style that incorporates some aspects of each of these.

Comment   In practice there is no single style of leadership which is effective for every situation likely to be encountered. An understanding of the merits and drawbacks of each leadership style will enable the most effective one to be adopted in each type of situation.

Sources   Chung, K.H., *Management: critical success factors*, Allyn Bacon, Neston, Mass., USA 1987.
Private communications.

Example   **6.4(b)**

Subject   **Demotivation of staff**

Background   Some years ago, after considerable discussion following cuts in government funding, a policy decision was taken by one UK Higher Education Institution (HEI) to leave the 'overheads element' of external research awards at the disposal of the academics concerned. In response to this incentive they trebled the HEI's total amount of overheads funds over the next three years and every academic development benefited either directly or indirectly. The amounts of money involved were very small compared with the annual budget of the HEI as a whole but the academics concerned appreciated the 'freedom' to spend this money at their own discretion. It was used for such purposes as:

- The funding of good postgraduates when other sources have been exhausted
- Doing speculative research before putting forward research proposals to funding agencies
- Bridging shortfalls in their running costs allowance
- Purchasing specialized equipment
- Contributing towards the building of more advanced experimental facilities

Some time later there was a threat of a small budget deficit during one financial period. This was the result of a combination of unreliable forecasts, too many simultaneous initiatives and overspending by certain sectors. The management's solution was to 'tax' the external research awards to cover the shortfall. At short notice and without prior consultation, a levy of 50 per cent was imposed on the overhead element of these awards. This action upset the long-term financial planning of many academics and aroused their bitter resentment. In consequence they were seriously demotivated in their efforts to attract new external research funding.

Comments   There are three common leadership failings that should be avoided in such situations. This case is a classic example of a decision that includes all three:

- *Lack of sound justification*:   There is no justification for such hasty action, especially when it reverses a previous, thoroughly debated decision that has been well supported by staff.
- *Lack of management judgement*:   It is very short-sighted to alter something that has been an important factor in staff motivation.
- *Lack of commerical logic*:   There is no commercial logic in antagonizing 30 per cent of the staff in order to gain access to resources which represent less than 0.25 per cent of the total annual budget.

Source   Private communication.

Example   **6.4(c)**

Subject   **Leadership**

Background   After retiring from his key role in ICI (Imperial Chemical Industries), one of the leading companies in the UK, Sir John Harvey-Jones wrote a best-selling book entitled *Making it happen—reflections on leadership*. In this book he has put down on paper many useful experiences that can assist those who aspire to be effective managers.

When he joined ICI it was a respected company which hired good graduates, had reasonably efficient management and was trading profitably in an established market, strong on chemicals. Sir John made his name by transforming ICI into an organization which is regarded as a performance leader in a very wide range of manufacturing areas including, for example, such products as paints and carpets.

In his book Sir John touches on a number of issues and some highlights are given here.

- *Winning support*:   He considered it of paramount importance for the leadership of an organization to involve the hearts and minds of those who have to carry out its policies and deliver the end result. This enables the entire workforce to understand the goal of the company and encourages them to give of their best to achieve it. As a result, in many instances members of the ICI workforce performed far better than would ever have been thought possible.
- *Stretching people*:   It is the responsibility of a good leader to give people the opportunity to develop as human beings. This can be achieved by careful planning, and making a variety of demands of them within their own environment.
- *Headroom*:   Once an objective has been agreed, people should be given

'headroom', i.e. scope for doing the work involved as they see fit, without continual interference from those in more senior positions.

- *Characteristics*: The prime characteristic of top leaders is toughness, both mental and physical. This is necessary because of the broad spectrum of severe demands that they have to cope with.

Comments   It is difficult to pinpoint the characteristics of the ideal leader because of the varied circumstances in which leadership is exercised. However, those with leadership experience can provide useful guidance to others on pitfalls to avoid and personal attributes to develop.

Source   Harvey-Jones, J., *Making it happen—reflections on leadership*, Fontana, London, 1988.

6.5 THE ROLE OF EDUCATION

Goal

To improve systematically the intellectual and moral faculties, expertise and skills of everyone involved in the activities of an organization.

About the subject

Education, in the broadest sense, has a key role to play in the success of any business activity, although the truth of this is not always fully recognized. As already indicated in Section 6.1, human beings are an organization's most valuable resource and constitute an investment that has to be continually enhanced if it is to offer a 'return on capital'. No matter how well qualified, however, new recruits cannot be expected to bring with them all the expertise needed to cope with every type of demand that could arise over a long period. This would be unrealistic in today's rapidly changing world. As products and services alter with changing demand, people have to adapt, alter or develop fresh expertise and skills to respond to new challenges. By making the educational enhancement of the staff an integral part of its business activity, an organization can expect to cope better with such changes and its policies are more likely to be readily accepted.

Apart from enhancing the professional skills of the staff, it is important to develop other faculties, such as the ability to differentiate between ethical and unethical choices and to recognize issues that require in-depth consideration. Individuals with identified potential should be encouraged to develop themselves as fully as possible. This is particularly important in international business activities because there is a need to understand and communicate with people of different languages, religions and ways of life.

Key questions and issues

Approaches to education

There are two basic views on the most effective approach to the education of technical personnel. One places emphasis on the need to acquire a sound understanding of scientific and engineering principles and leaves little time for demonstrating how these are employed in practice. To support this view, it is claimed that with a solid foundation further development can readily be achieved as experience is gained during the person's career. The other view is that technical personnel should have an appreciation of engineering principles, but that it is more important for them to be taught the most up-to-date knowledge and techniques so that they can contribute positively once they take up employment.

Both approaches have strong supporters, but organizations need to strive for a balance between the two. There is no doubt that technical personnel should have a good understanding of the basic scientific and engineering principles, but they must also have the confidence gained from practice in applying these principles to real problems. Technical personnel also need to have an understanding of other matters such as business topics, communication skills, psychology, organizational behaviour and social responsibility.

The main approaches in higher education

For a number of reasons, we are unlikely to find the ideal educational programme to satisfy everyone. Typical examples of the kind of difficulty encountered include the question of whether an educational programme should aim for breadth of knowledge or concentrate on a few specific topics so that sufficient depth can be reached in these. A broad understanding of a range of topics is highly desirable because it widens the outlook, but there is the problem of integrating new knowledge into this framework so that it can be applied effectively. Also, since only a limited amount of teaching time is available, it could mean that only a superficial understanding of any aspect is acquired.

On the other hand, producing highly-specialized engineers would have its own advantages and drawbacks. Engineers with high technical competence will certainly make a useful contribution to their organization. However, technologies change rapidly and unless these engineers are mentally and technologically prepared for change they may lack the flexibility to tackle new challenges effectively. The principal goal in higher education must be to achieve a balanced programme that prepares graduates to cope with new challenges as they arise.

A *way forward*

Following the points made above, there is clearly a need to develop an educational approach that can overcome the problems identified. The starting point must be to ensure that the desired basic requirements for an engineer will be developed, and these have been indicated in Section 6.1 as a balance of competence, confidence and communication skills.

Such an approach has been generalized under the title of 'The 3-C Educational Approach' and it would go a long way towards meeting the demands of a versatile and flexible educational progamme. Figure 6.8 illustrates how such an approach can be put into practice.

This approach allows a fundamental understanding of a wide range of subjects to be acquired so that their relative importance is fully appreciated. The curriculum should include science, business, social and communication skills alongside the engineering subjects that interface with these topics. Core engineering subjects, however, should receive the most attention to ensure that students achieve technical and professional competence. Only a selected number of topics in each of the main subject areas are then taken to greater depths.

The main purposes of this approach are:

- To train students in taking the study of a subject right through from initial appreciation to thorough understanding so that they gain confidence in their learning skills.
- To give them experience in using the same basic methodology for tackling other subjects with their own special assumptions and terminologies.

In this way, only a limited amount of knowledge is 'stored in the head', the emphasis being on methods of learning and reasoning, while additional information can be obtained as required from libraries or databases.

Educational enhancement

The management of an organization has to decide how the educational enhancement of its staff is to be achieved.

One approach is to opt for strategically selected programmes closely linked to job requirements. These would include day release classes on a weekly basis, evening classes outside normal working hours, in-house courses, or intermittent blocks of special courses. Alternatively, staff can take part in programmes which are available on a 'free-standing' basis but involve attendance at a specific location for an extended period. Typical examples would be courses at colleges and universities, and specialized intensive programmes lasting from two days to two weeks.

In practice, different people respond better to different approaches. Some find the first

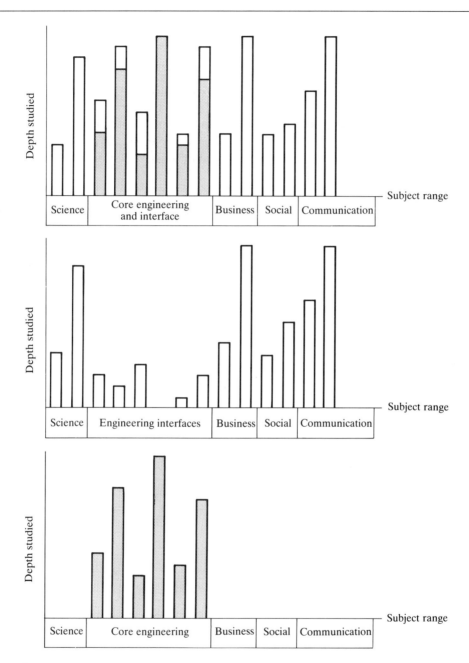

Figure 6.8 Distribution of subject range against depth studied

approach easier to handle while others prefer and benefit more from the second. The decision in each case is up to the judgement of management, but two points need to be recognized.

The first is the need to include a specific period for educational enhancement in, say, the annual work schedule for every member of staff. This arrangement should be strictly adhered to, as there is never a 'good' time to release key personnel for training purposes. The second is that some method should be devised for measuring and rewarding the benefit of such training periods so as to encourage enthusiastic support from everyone.

Planning courses

Courses are now available from a range of sources. Some, based on the findings of market research, are offered by educational institutions. Others are organized by companies themselves to meet specific needs at a given point in time for their staff. Still others have been devised by commercial organizations to meet the demand for upgrading of staff in industry.

Ideally, courses should be devised jointly by clients and suppliers, and this is particularly relevant to in-service training courses. The clients should be able to specify what they require while the suppliers indicate what can be effectively achieved, and the best way to go about meeting the need. The basic questions to be asked will include the following:

- Does the course meet the objectives of the organization?
- How well has it been planned?
- What teaching methods are employed, and are they effective?
- Do the providers have the necessary teaching experience and academic standing?
- What is the quality of the material provided for participants?
- What is the track record of the course?

By seeking satisfactory answers to these questions clients are more likely to obtain what they want from the courses.

Practical difficulties

There are a number of practical difficulties which should be recognized by organizations seeking the professional development of their employees.

Firstly, the effort of preparing a specialized course might not be justified on a commercial basis if the requirement is too specialized and the course would be useful to only a very small number of potential participants.

Secondly, courses for in-service training tend to become out of date very quickly, and it has been estimated that the average 'life' of such a course is only 18 months. Considerable effort is required to ensure that the best use is made of prepared teaching materials while they are still relevant.

Thirdly, it is difficult to cover a number of diverse requirements within a short space of time, such as a few days. As a result, many clients have found even the most widely advertised courses insufficiently specific for their needs.

There is no ideal solution to these problems, but one realistic approach would be to identify the various 'suppliers', prioritize the types of course offered in relation to the organization's needs, and select courses for individuals or groups on the basis of how well they meet specific objectives.

ILLUSTRATIVE EXAMPLES

Example   **6.5(a)**

Subject   **Short courses for professional updating**

Background   As technologies continue to develop, a useful way for career engineers to keep themselves up to date is to attend short courses on relevant technological and managerial topics. There is a healthy demand for such courses and two examples will now be considered.

● *Continuing education courses*:   The principal aim of these courses is to update professional people, e.g. practising engineers, on developments in their own field. Programmes are organized by academic institutions or organizations closely linked to such an institution and are given by experts. Each one covers a specific range of topics and is highly concentrated. Courses normally last between three and five days and take place either in the educational institution responsible or in a hotel or other suitable location.

Typical of such programmes are those offered by the Center for Professional Advancement which offers over four hundred courses in the US and a hundred in Europe at a unit cost equivalent to three working days' salary for a young engineer. Courses offered for personnel from the oil and gas industry include: Contracting and Contract Management; Cost and Planning Economics; Gas Turbine Technology; Documentation and Quality Assurance; and Valve Technology.

● *Commercially organized courses*:   Integrated Computer Systems is a company that regularly offers a range of courses to meet various requirements. There is, for example, a series of nine programmes under the title of 'Courses for Managers in Technical Environments'. Each lasts either three or four days at a unit cost equivalent to four working days' salary for a young engineer and they are offered at four locations in the USA, two in Canada and three in Europe. The European locations offer the attractions of Paris, the culture of London or the sophistication of Stockholm! The topics covered include Management Skills, Finance, Project Management, Planning and Interpersonal Skills and 'purpose-built' course material is also provided. The publicity brochures are well presented, highlighting special features of the courses as well as their content, and giving clear information about how to make contact with the organizers.

Comments

The success of any particular course depends on the way participants respond to the teaching methods used, and on the relevance of the material covered.

Courses arranged by commercial organizations are generally presented more attractively. Academic and professional institutions, however, offer programmes that lead to some kind of additional qualification, which is often very appealing.

Sources

Leaflets provided by Integrated Computer Systems and the Center for Professional Advancement, USA.
Press comments and private communications.

Example

6.5(b)

Subject

Training centre for engineers

Background

After twleve years of discussion and planning a technology training centre called the Indian Metrology Centre was set up in the National Physical Laboratory in New Delhi, India. The main task of this Centre is to train engineers to understand the importance of accuracy in industrial activities. This understanding is particularly relevant to the following aspects of industry in which, either directly or indirectly, accuracy plays a significant part:

- Industrial safety
- Product reliability
- Standardization
- Manufacturing efficiency
- Implementation of new technologies

For a developing nation these are all matters of vital importance. The first one has particular significance for India, because it was at Bhopal in 1984 that poor standards of industrial safety resulted in a major disaster, the memory and effects of which are still very real for those who were affected by the poisonous gas which leaked from a pesticide plant. The remaining four are important for all developing nations wishing to compete in the world market for manufactured goods. If sufficient attention is to be given to accuracy in all these five aspects of industry it is essential to have an adequate number of engineers with expertise and experience in metrology.

An institution like the Indian Metrology Centre will provide engineers with training that will enable them to serve their country's industries in a much-needed area of expertise.

Comment   Educational institutions such as the Indian Metrology Centre may not have the glamour usually attached to activities associated with computer technology, electronics or artificial intelligence, but their contribution will have a fundamental impact on national development and prosperity.

Source   Land, T., Training centres help poorer countries to compete, *Financial Times*, 27 July 1989.

Example   **6.5(c)**

Subject   **Achieving professional chartered status**

Background   After completing their studies and obtaining initial qualifications one way for engineers to enhance their development is to acquire recognized professional status while gaining practical experience on the job. In the UK such professional status is usually indicated by the designation 'Chartered Engineer (CEng.)' while in North America the equivalent designation is 'Professional Engineer (PE)'.

A typical example of how to acquire such professional status is given in the booklet produced in 1987 by the Institute of Civil Engineers in the UK, and entitled 'The Routes to Corporate Membership'. Its key points are summarized as follows:

> The engineer has to satisfy the Institute at two levels of experience referred to as Professional Examination Part 1 (PE1) and Professional Examination Part 2 (PE2) before he can become a corporate member of the Institute and hence have a recognized professional status.

The objective of PE1 is to assess the 'quality of a candidate's practical experience of civil engineering and to test his technical competence'. The examination of the candidate will cover, for example:

- Grasp of the application of engineering principles
- Ability to apply standard procedures and techniques
- Ability to assemble facts, thoughts and ideas logically and to communicate effectively
- Ability to assess the financial implications of activities

PE2 has a similar objective, but uses different criteria to judge professional competence. These would include:

- Sound scientific knowledge
- Good engineering judgement

- A responsible attitude towards safety
- The ability to put his or her knowledge to creative and imaginative use

In each case specific requirements of experience must be fulfilled before a candidate is eligible for the examination and these are clearly defined.

Comments   There is an incentive for both young engineers and their employers to make a positive effort to ensure that they achieve a recognized professional status. For the candidate, it is a recognized qualification that is valid throughout the world wherever he or she may wish to work. For employers the educational enhancement involved increases the effectiveness of the engineer's work. On a more general level it provides a recognized method of enhancing the quality of an organization's entire engineering staff and hence its competitiveness in the market-place. In a bid for a contract, for example, the quality of its staff may well be a key factor in determining the company's success or failure.

Source   'The routes to corporate membership', The Institution of Civil Engineers, UK, *ICE*, No. 43, 1987.

6.6 ORGANIZING TIME

Goal

To make the most effective use of time.

About the subject

It is many years since the Americans coined the expression 'Time is money' and the sentiment is still misunderstood in other countries as one of simply putting money above everything else. Recently, however, the value of time has begun to be more fully appreciated by everyone. In business activities, time has to be regarded as a resource, and one which is directly related to human performance. It is, therefore, the starting point in pricing any contract work. A typical project, for instance, may require X number of hours of work by Grade A personnel, and Y number of hours by Grade B personnel. Each grade has a different value. The rate is expressed in this case as 'cost per hour', and the total estimated time and cost for the work may have to be adjusted to allow for variations in the availability of suitably qualified staff. In calculating the amount of work time available for specific tasks over a period of one year, allowance has to be made for annual leave and public holidays.

Since the amount of time at our disposal is finite, we need to identify the different demands on it and develop suitable methods for its allocation. For an engineer, typical demands include time for doing project work, attending meetings, travelling, participating in conferences, following leisure pursuits and socializing.

Key questions and issues

Activities needing a time allocation

Every engineer needs to allocate blocks of time for taking part in some of the following activities:

- The technical aspects of work assignments
- Attending meetings both within and outside the organization
- Travelling to meetings, conferences and presentations
- Administrative duties
- Preparing reports and making presentations
- Discussion with colleagues, clients, visitors and collaborators

Outside his work situation, time has to be allocated to at least some of the following:

- Travel to and from work
- Recreational activity
- Socializing with friends
- Family concerns and household administration

It is important to know how much time is being devoted to these various activities in order to assess the best use of the hours available.

It also needs to be recognized that time allocations are not static but will need constant revision as one's career develops.

Time planning

The first stage in planning the use of time is to gather information about how it is presently being used. This is best done on a weekly basis over a four-week period, with the aid of a time planning chart. This may be divided into one-hour, half-hour or quarter-hour blocks. Figure 6.9 shows a typical time allocation chart in which each typical function is assigned a code number to facilitate daily summing up. By the end of the selected period it will be possible to calculate fairly accurately the average percentage of time devoted to each function.

This chart can also be used to determine whether there is a proper mix of the various functions and to help identify where excessive time has been spent on one type of

Name of person: Week ending: / /19

			Time																	Time total per day (hrs)						
Day No.	Date	Day	7	8	9	10	11	12	13	14	15	16	17	18	19	20	21	22	A	B	C	D	E	F	Total Hrs.
1		Sun																							15
2		Mon																							15
3		Tue																							15
4		Wed																							15
5		Thu																							15
6		Fri																							15
7		Sat																							15
																			Total						105
																			Percentage						100%

A: Professional activities   D: Meals and entertainment
B: Meetings   E: Social and recreational
C: Travelling and visits   F: Miscellaneous

Figure 6.9   A data sheet for measuring time allocation

activity, such as work at the computer terminal or attending meetings. Such information will enable one to begin apportioning time much more effectively.

Organization of time

Once the initial survey has been done, the most likely 'discovery' is either that there is insufficient time for all that has to be done or too much is being spent on work which cannot be adequately justified or on matters which would be better delegated to others more directly responsible.

One response to this problem is to carry work tasks and responsibilities into the hours beyond the normal working day, and this has resulted in some senior managers working for up to 60 hours per week. Much more effective would be to prepare a list of all one's functions in order of importance so as to assess what must be done immediately and what can be deferred or delegated. This would ensure that each item of interest is systematically noted and its time allocation considered. For someone in a senior position, time has to be found for regular fact-finding tours of the organization—such as visits to the shop-floor or different factories, individual discussions with key personnel on specific problems, and briefing sessions for staff on the company's prospects and business performance.

Project engineers need to allow time for updating themselves on technological advances such as the application of more efficient computer software, and gaining practical experience to support theoretical work on topics.

Handling several tasks concurrently

Very few technical personnel have the opportunity to concentrate on a single assignment, and nowadays they are often expected to contribute to several concurrent projects. This is done by membership of, say, two project teams or on a consultancy basis. This state of affairs is quite logical. Although the same specialized expertise may be needed on two projects, possibly funded from different sources, it is unlikely that either will require the full-time commitment of the person concerned. The best use of his or her time, therefore, is to allocate an agreed proportion to each project as required.

Some technical personnel get concerned at being asked to take on several assignments at the same time. In practice, such an arrangement can be highly beneficial, provided proper time planning and training are introduced. Handling two different tasks concurrently may well prevent 'staleness' developing through over-concentration on the problems of a single assignment, and can often result in a fresh approach to both.

There are, of course, inherent dangers if such a practice is carried too far, and beyond the capability of the individuals concerned. If this happens, performance in all the projects will suffer because deadlines cannot be met or because the time required has been underestimated.

Time and work efficiency

Efficiency is closely linked with how one uses time and approaches work, and these factors are, in their turn, related to leadership and team activities. Using the term 'delegation' in the sense of both the allotment of tasks to others by a leader and the sharing out of the work of an assignment by team members, it is possible to analyse someone's utilization of time with the capacity to delegate.

- *Efficient delegation*: The organizer delegates to colleagues responsibility for different aspects of a given task and leaves them free to do the work as they find best within the time available. The results are then collated at the agreed deadline. This approach makes the most efficient use of available time by limiting the amount of work required of each individual and yet facilitates punctual achievement of the objective of the task.
- *Inefficient delegation*: The organizer systemtically delegates, but voluntarily or involuntarily takes over responsibilties which actually 'belong' to colleagues. As a result, the organizer becomes increasingly short of time and inefficient.
- *Inadequate delegation*: The organizer is overwhelmed with commitments

because of a wish to be involved in every activity. As a result all decisions have to be processed through the organizer, whose own volume of work becomes too high. Even more than in the previous case this leads to a general shortage of time and extreme inefficiency.

Assigning priorities

The first step in organizing a series of the tasks is to assign to each an appropriate level of priority. Obviously, those with the highest priority rating must be dealt with first, either by the one immediately concerned or through delegation. The corporate plan of an organization, for example, is something that can be prepared either by the chief executive or by a team of colleagues. The ability to prioritize tasks and to estimate the time required to perform each one is important for any professional, but a high degree of proficiency in this is only acquired through practice and experience.

It must also be recognized that priority ratings and the time available for different tasks will alter as a career develops and different types of appointment are undertaken. A young engineer in his or her first job, for example, should be devoting all energy and attention to the specific task which has to be carried out. With eventual promotion to Section Leader, however, comes the need for delegation in order to be able to cope with the additional responsibilities.

ILLUSTRATIVE EXAMPLES

Example   **6.6(a)**

Subject   **Contrasting approaches to the use of time**

Background   Here for consideration are two instances of very different approaches to the use of time.

The project leader: A leader of a project team was found to be typing the handwritten draft of a final contract report on a wordprocessor and editing it himself before passing the material to a very competent secretary for typing into the submission version of the report. His reasons for this were twofold. Firstly, he found it more convenient to make corrections to the technical materials on a wordprocessor. Secondly, he believed that the typed version could be more readily copied by the secretary in the agreed format for the final report.

Calculations show that the project leader's hourly pay rate was 2.5 times that of the secretary and the secretary's speed of inputting information on the wordprocessor was four times that of the project leader. In other words, there was a 10:1 inefficiency factor.

By the time this state of affairs was discovered, the contract report was

running well behind schedule and quick action had to be taken to cut out
the middle phase (the project leader's typing). Giving the secretary access
to the handwritten draft for typing before the project leader made
corrections to it helped greatly in meeting the deadline.

The chairman of a company: The chairman of an electronics company
decided to map out a time schedule each January with the help of a small
team. They were to allocate suitable portions of his time over a six-month
period to each of his various activities. This arrangement would enable his
colleagues to plan their programmes as well.

In 1990, for example, he decided to devote one quarter of his time to
visiting the company's customers and spend 60 per cent of his lunch
periods on public relations activities. Weekends were used for thinking
over important issues. This was often done subconsciously while playing a
round of golf! It was his hope that the result of this approach would be
more time for other activities and also more effective delegation.

Comments   In the first instance, we have a capable technical person whose priorities
are all wrong. He was more interested in his own method of doing this job
than the efficient use of his time. It was much more efficient for the
handwritten draft to go direct to the typist. He had also disregarded the
time schedule agreed for the project.

The second instance shows that anyone in a senior position has to be firm
about time utilization, otherwise it will be impossible to fulfil his own
priorities. It may seem inflexible to allocate one's use of time so far in
advance, but it helps others to plan the efficient use of their own time as
well.

Sources   Private communications.
Press comments.

Example   **6.6(b)**

Subject   **The effective use of time**

Background   Effectiveness in the use of time is not always directly related to the
amount spent on a given assignment and this is particularly true in
business activities. Two typical illustrations will highlight the truth of this
point.

The technical manager of an engineering company: This manager is on
the board of a busy engineering company and puts in around 30 per cent
more work time than the official schedule. The recognized working day for

all staff members is from 8.20 a.m. to 5.00 p.m. with half an hour for lunch, but his typical day lasts from 7.00 a.m. to 6.00 p.m. He also takes work home in the evenings and some weekends. An analysis of his time usage shows the following distribution:

- For meetings: 50 per cent
- On telephone: 20 per cent
- Travelling: 20 per cent
- Other activities: 10 per cent

The final element in the above allocation is the time used for reading, paperwork, preparation for meetings, planning and thinking.

Time spent with a decision maker: The research manager of a consultancy put forward a carefully prepared project proposal to a major oil company in the hope of receiving support funding. The proposal dealt with improved methods of oil production and offered potentially attractive cost-saving benefits. It was well received by the various sections of the organization. However, there were several questions to be answered before the company would give a decision.

After several months' delay, and interaction with various members of the oil company staff, the research manager decided to seek a meeting with the decision maker on research and development. This would take the form of a short presentation which would be attended by a number of other key personnel. Careful preparation was made for this half-hour meeting and questions posed during it were satisfactorily answered. The outcome was that the company gave a favourable decision on the proposal immediately thereafter.

Comments   In the first case there is clearly insufficient time for private thinking, long-term planning and exploring new market opportunities. A revision of this technical manager's activity distribution to give more time for these activities might well lead to greater effectiveness and efficiency on his part.

In the second case, discussion of unclear factors had been protracted to a point where it seemed no decision would be forthcoming. Yet once the consultancy's research manager was able to arrange a comparatively short meeting with the company's decision maker the difficulties were quickly overcome and the proposal was immediately accepted for funding.

Source   Private communications.

Example	**6.6(c)**
Subject	**Time organization**
Background	The term 'busy' is usually employed to express the sentiment that in a given period of time a large number of activities have to be carried out. It is usually difficult to quantify what is meant by 'busy-ness' but there are situations where this is possible and it can be readily calculated.

During a flight between Brussels and London on a well established airline the technical manager of a company specializing in heavy engineering had just such an opportunity. A request to a member of the cabin staff at the beginning of the flight was not responded to until it was repeated towards the end of the journey. The reason given for this non-response was that the crew were 'very busy'. In this case the 'busy factor' can be quantified as follows:

$$\text{Busy factor} = \frac{\text{Passenger load}}{\text{Crew availability}}$$

$$= \frac{(\text{no. of passengers})/(\text{no. of seats})}{(\text{no of staff on duty})/(\text{full crew for flight})}$$

Clearly, a fully-loaded plane with a full complement of crew will yield a factor of 1.00, since both ratios would be 1.00. Normally any value above 0.8 would be classified as 'busy'. However, on this particular flight the plane was only half-full and there was a full complement of crew on duty. The 'busy' factor in this case was equal to 0.5, which could hardly be considered 'very' busy!

Comments	Attempts should be made to identify criteria for measuring the utilization of time. It is all too easy to use the term 'busy' to describe poor usage of time, inefficiency or simply an unwillingness to be helpful.
Source	Private communication.

6.7 MATERIAL FOR FURTHER STUDY

Adair, J., *Effective team building*, Gower, Aldershot, 1986.
Adair, J., *How to manage your time*, Talbot Adair, Guildford, 1987.
Adair, J., *Developing leaders*, McGraw-Hill, Maidenhead, 1988.
Beer, M., *Lead to succeed*, Mercury, London, 1989.
Bittel, L., *The McGraw-Hill 36 human management cases*, McGraw-Hill, Maidenhead, 1988.

Goodworth, C., *Effective interviewing*, Hutchinson, London, 1987.

Kuo, C., *The 3-C approach to engineering education*, World Conference on Engineering Education for Advancing Technology (WEEAT), February 1989.

LeBoeuf, M., *How to motivate people*, Sidgwick and Jackson, London, 1986.

McKenna, E., *Psychology in Business*, Lawrence Erlbaum Ass., London, 1988.

Martin, A.S. and Grover, F., *Managing people*, Thomas Telford, London, 1989.

Maslow, A.T., *Motivation and personality*, Harper and Row, London, 1954.

Rawlinson, J.G., *Creative thinking and brainstorming*, Wildwood House, Aldershot, 1986.

Reynolds, H. and Tramel, M.E., *Human behaviour in organization*, Pitman, London, 1983.

Taylor, R.L. and Rosenbach, W.E., *Leadership challenges for today's manager*, McGraw-Hill, Maidenhead, 1989.

CHAPTER 7

CASE EXAMPLES

7.1 INTRODUCTION

In the foregoing chapters the fundamental elements of markets, management, money and manpower have been individually examined and illustrative examples have been used to highlight specific aspects of each topic in turn. Fuller case examples will now be considered to demonstrate how the integration of these four elements affects an organization in practice, highlighting their interdependence and the contribution they make to the success or failure of specific operations. Further useful lessons will also be drawn from the cases cited.

Each case will be examined under the following main headings:

- Background
- Basic questions
- Case data
- Performance analysis
- Comments
- Present status

7.2 AN OIL COMPANY

Background

This example is concerned with Burmah Oil plc and is based on the lecture given by L.M. Urquhart, chief executive, on 24 October 1989 at the Annual Meeting of the West of Scotland Branch of the Institute of Petroleum. Urquhart's paper was entitled 'Burmah Oil: management lessons from two decades of turbulence', and reviewed the company's experience during the whole period 1960 to 1989, but concentrated on the five years between 1969 and 1974.

Basic questions

From the background a number of basic questions can be asked:

- Are you familiar with this company?
- If so, did you think it was a conventional oil company or an industrial company?
- If not, what would the name of the company suggest to you about its activities?
- Why did the speaker concentrate on the period 1969 to 1974 and what key event occurred during that period?
- Do you think this is a success story, a failure lesson, or a combination of the two?

It would be interesting to write down your answers on a separate sheet of paper and compare them with what follows.

Case data

Burmah Oil was founded in 1886 and initially its main activities were oil exploration and exploitation in Burma and India. By the late 1960s, the Burmah Oil Group had become virtually an investment trust, deriving over 65 per cent of its income from its 23 per cent investment in British Petroleum (BP) and its 3 per cent holding in Shell. However, the dividend generated from these investments was 'a very poor return on capital invested' and had to be passed directly to the company's shareholders. Eventually, in the light of the widespread view within the group that better use could be made of its resources, it was decided to diversify into areas away from its core activities, and the group embarked on a £450 million programme over the period 1969 to 1974. This comprised three key initiatives.

Initiative 1

Various enterprises were acquired 'downstream' from oil production itself. These ranged from a by-product user in Rawlplug (the industrial and do-it-yourself fixings and manufacturer) to companies with petroleum-related interests such as petrol station chains, bicycle and car accessories shops, a crane company, a luxury caravan manufacturer and a major manufacturer of car parts.

Initiative 2

A second area of involvement was shipping. In 1971 a company was set up in New York to ship crude oil to Burmah's refinery at Ellesmere Port (UK) and to operate a tanker chartering business. By 1973, the company had 21 vessels under its own flag and on charter. Burmah's board was considering and approving plans to establish a transshipment terminal in the Bahamas costing over $30 million and further shipping was commissioned to enable the company to offer an integrated transport package. By 1974 there were 42 tankers on the books. In October 1972, the board sought and was awarded a contract to ship liquefied natural gas (LNG) from Algeria to the eastern United States. This involved the purchase of three new LNG carriers, which at the time were regarded as the most expensive type of ship to build. Within a year, a further project was effectively in place for the transportation of LNG from Indonesia to Japan, and five more vessels were ordered.

Initiative 3

In January 1974 the Signal Oil and Gas Company of the USA was acquired for $480 million with the help of a loan from the Orion and Chase Manhattan Banks of USA on the expectation that Burmah's BP shares would be profitably sold after its own shareholders had been consulted. This acquisition gave Burmah Oil a stake in the North Sea since Signal had recorded a successful strike in the Ninian Field in 1973. However, the capital needed to develop this strike was estimated at £450 million—far more than the much larger BP was spending on any single capital project.

Performance analysis

In February 1974, Burmah Oil had a market capitalization of £700 million and was seen by the investment community as a 'blue chip' (a very highly regarded) company. By February 1975, its market capitalization had fallen to £55 million, its share price had dropped by 92 per cent and its earnings per share went from 31 pence in 1973 to a loss of 17 pence in 1975. It would be useful to analyse the performance of this company during the period 1969–74 from the point of view of the four M elements: markets, management, money and manpower.

Markets

Burmah Oil changed itself from an investment company to a conglomerate in a very short space of time by entering a number of new areas of activity in which it had little or no expertise. It can be said, however, that the board ignored the principles of marketing and its market research was less than rigorous. It is not clear whether the members gave serious consideration to the types of business in which to become involved, or asked whether it was possible to integrate oil exploration and production with automotive retailing, running a fleet of 40 tankers and managing the complex business of LNG transportation.

If a company decides to expand its areas of activity, it must ask several key questions and be satisfied with the quality of the answers obtained. These questions should include:

- How should perceived opportunities be prioritized?
- How reliable are available market forecasts?
- When would be the correct time to act?
- What is the present level of competition?
- What alternatives are already available?

It seems likely that the Burmah board paid insufficient attention to these points and that its decisions were based solely on the most optimistic forecasts.

Management

Any organization undertaking a major change of direction must have clearly defined objectives and credible strategies. The board of Burmah Oil appears to have been without either. Developing a credible strategy calls for a proper identification of an organization's strengths and limitations. Successful implementation of a strategy will involve maximizing the use of its strengths and minimizing the demand on skills which are not readily available. Where strategies are formulated their robustness must be checked by considering both the best and worst possible scenarios while taking into account likely political and economic changes.

The handling of contracts also appears to have been weak. Questions arise about the flexibility of both those concerned with operating tankers and those involved with the borrowing of money. In addition, it can be said that it was unwise to rely too much on Burmah Oil's shareholding in BP in drawing up the latter contracts.

On another aspect of finance, Urquhart suggested that Burmah Oil paid too much for many of its acquisitions at that time, and this was borne out by the quotation from a leading newspaper: 'Bids, Oil Company bids, and Burmah bids'. Good negotiating skill can usually achieve effective deals without undue financial outlay.

In the case of all three initiatives, carefully conducted feasibility studies on the various proposals would have cast serious doubts on the viability of many of them. Equally, if the management had sought more relevant information at that stage and gained a fuller knowledge of the performance of the companies concerned it would have been able to take appropriate action to avoid the potential problems identified.

Money

The first oil crisis of 1973 took the industrial world by surprise and as a consequence every company suffered. Some knowledge of the fundamentals of economics and government policies at Burmah Oil would have been helpful at that time, and might have reduced the impact of the crisis on company operations.

A well-managed organization must also have sound accounting methods, a proper reporting procedure and effective financial control. This point is best illustrated in regard to Burmah's marine activities. The shipping operations were conducted as 'an almost entirely autonomous unit' at arm's length in New York with little supervision and this office did not even produce 'proper management accounts'. It is often felt that the need to prepare budgets, conduct strategic reviews, and submit valid proposals etc. are bureaucratic nuisances. However, a healthy management will want to have such information available for a variety of reasons. It provides a record for future reference, it can be the basis of a performance appraisal, and it may be drawn on when a second opinion is sought on a problematical situation.

Manpower

The quality of the personnel in any organization is difficult to judge from the outside, although there are always some indicators if they can be picked up. The basic problem in such judgement stems from the need to balance two different requirements. On the one hand, an organization must have people with entrepreneurial flair if it is to operate successfully in a highly competitive market. On the other hand, it needs a measure of rules, procedures and structure to control activities. Too much of the first characteristic, however, can be damaging to its operations, and too rigid a control may result in under-performance despite a steady income. At the same time the organization would probably fail to attract personnel of outstanding quality. In the early 1970s the Burmah board allowed itself to be carried away by the enthusiasms of some of its more energetic members into areas where caution would have been a better policy. The company was obviously happy to support those who could turn an investment company into a dynamic organization which could play a leading role in the commercial world. However, the enthusiasm of this group dominated the redirection and reorientation of the company and other people went along with their initiatives without questioning either the merits or the justifications of the latter.

Commenting on this factor, Urquhart emphasized the need to recruit 'high-grade personnel of independent mind'. Although he did not define the term 'high-grade' it was clear he attached great importance to the quality of an organization's human resources. All the indications are that the characteristics he had in mind are:

- Professional competence
- Communication skills
- Judgemental skill and the ability to prioritize demands
- The readiness and the ability to operate either autonomously or under supervision
- An independent mind

In Burmah Oil's case some of these qualities were not available in sufficiently large quantity at the higher levels of management.

Comments

While the company was changing itself from an investment group to a conglomerate four significant and related events were taking place.

- The Yom Kippur war, in October 1973, followed by the quadrupling of oil prices
- A sudden drop in the amount of crude oil to be transported
- A sharp downturn in the world economy
- The 'winter of discontent' in Britain with a three-day week in operation

The immediate consequence for Burmah Oil was a 60 per cent drop in the value of its shareholding in BP which created a need to renegotiate the loans for the purchase of Signal Oil. At the same time the tanker fleet was without adequate business and was losing money fast. In order to survive, the group had to appeal to the Bank of England for help.

From the well-documented information available it is clear that the senior management at Burmah Oil over this period lacked a balance of the four Ms and also had serious weaknesses in certain aspects of each M, which contributed to the near-bankruptcy in 1974. The major deficiencies are as follows:

- *Management*:   A lack of credible strategy for fulfilling the group's ambitions. A lack of control by management of the group's activities. Poor use of negotiating skills to get the best deals.
- *Markets*:   There was no logic in many of the acquisitions made, and this led to an unnecessary expenditure or effort on integrating new areas of business for which the group lacked adequate experitse.
- *Money*:   Proper control of finance was not exercised and an adequate understanding of economics and government policies was lacking.
- *Manpower*:   The enthusiasm and energy of key people were not matched by good leadership.

Present status

During the 1980s Burmah Oil was completely transformed from an oil exploration and exploitation company to a company with three core interests:

- Lubricants
- Special chemicals
- LNG transportation

The board's strategy has been to concentrate on those areas in which the company had undoubted expertise, to strengthen the balance sheet and to eliminate activities which management does not fully understand and which are making an unsatisfactory return on capital invested. As a result, the company has returned to profitability and is reinforcing its reputation as a high-quality company.

7.3 A PRODUCT IDENTIFICATION COMPANY

Background

This second example considers a small but fast-growing company which concentrated on providing high-quality information labels. The company, Donprint Label Systems, was founded by Des Donohoe and its factory is at East Kilbride, a town in Scotland near Glasgow. Its labels may be found at the back of equipment supplied by computer companies, such as IBM, Honeywell, Wang and Digital, giving key instructions for users including voltage and connection requirements. The material used here was obtained through personal contacts and from the company's brochure and other published literature.

Basic questions

- Do you think printing labels is a business with potential?
- Do you believe this is a niche market?
- How important do you think quality assurance is in this business?
- Does this business need a lot of funds for its operations?
- How many people would you expect to be working in the company?

Once again, write down your answers on a separate sheet and compare your views with the text.

Case data

The company was started in 1979 with a capital of £500 which was used to buy a second-hand labelling machine and to take an option on a thousand square feet of factory space. In classic entrepreneurial fashion, the founder went out during the day to seek orders and worked in the factory at night to meet them. Turnover in the first year was £20 000 and there was no profit. This did not deter Mr Donohoe, however, as he believed that as long as he gave customers what they wanted when they wanted it, his business would flourish. By 1986 Donprint had a total workforce of 26 and turnover was up to £1.5 million. The company's core business had been established in the field of product identification technology.

Performance analysis

Performance over the first ten years can be examined under the headings of the fundamental elements.

Markets

Marketing strategy is based on the decision to focus sharply on clearly defined market sectors so as to become the best supplier in certain specific areas. The development of this strategy has led to the establishment of over four hundred quality or 'blue chip' customer accounts in the United Kingdom and Europe. These include many multi-national companies in the following industries:

- Computers
- Electronics
- Electrical
- Telecommunications
- Domestic equipment
- Automotives
- Aerospace

These customers demand the highest quality and, in order to retain their accounts, Donprint has to maintain its standards and its competitive position. This requires the company to be at the leading edge of product identification technology. It also implies the need to satisfy a number of quality assurance standards and procedures such as:

- British Standards 5750 Part II
- ISO 9002
- 'Acceptance' of the Canadian Standards Association

To develop the customer base, effective selling skills had to be developed and credibility established through building up the necessary knowledge to satisfy customers' requirements. Extensive market research had also to be carried out on a regular basis in order to identify changes in needs. To stay in the forefront the company had to innovate with new materials, new approaches and new applications.

Management

The management team had a clearly defined objective. This was to provide the best products and customer service in the field of product identification within a small number of market sectors.

The strategy for meeting this objective is based on a total commitment to quality and excellence in all activities. In practice, its implementation calls for effective project management, good contracts and positive negotiation with a wide range of customers.

For planning and future development purposes the management has recognized the importance of having an effective information management system. For example, everything associated with its product identification technology has to be documented so that its labels can be reproduced exactly in other countries and languages. To ensure that the company keeps ahead of competition, research and development efforts have been applied to the design of 'future solutions'.

Money

During the first decade, the company expanded very rapidly and financial resources had to be found to support the implementation of the management's strategy. The management struck a careful balance between having ease of access to the required finance and maintaining sufficient independence to be able to meet its own objective. The majority of the required funds were, in fact, generated from revenue earned in the operation of the business. Once the necessary finance was obtained its utilization was very carefully monitored. The policy adopted was to have tight control on finance, the use of resources and the meeting of production deadlines. In addition, the viability of new developments was rigorously assessed in the context of a policy of active response to the needs of customers. Government policies were carefully studied in order to take up opportunities offered and to be able to overcome any constraints they might impose on the company's activities.

Manpower

To support its ambitious programme, Donprint requires high quality staff and the task of finding the right people has become very difficult because of the management's demanding criteria. It looks for recruits who are professionally competent in the necessary areas of new technology and who are at the same time fully committed to the philosophy of the company. The problem is increased by the fact that new industries are developing in the East Kilbride area, and the demand for skilled human resources has risen sharply.

The management's response to this situation has been to adopt a policy of encouraging employees to enhance their skills and knowledge. As a result, around 30 per cent of the staff are on some strategically planned programme, such as a three-year day-release course, or a two-year programme of night classes. The company identifies suitable courses, pays the tuition fees and, where necessary, adjusts work schedules to facilitate staff members' attendance. At the same time a further 20–25 per cent of the staff are on courses which assist personal development and increase their potential benefit to the company. The most popular of these courses are those associated with languages and commercial topics.

Comments

The main reason for the success of Donprint has been the management's firm grasp of the importance of covering all four Ms. Close examination of its policies and evaluation of its progress clearly demonstrate this point.

Markets

The marketing policy is based on complete dedication to the needs of customers and a total commitment to quality and excellence. In addition, the company set itself the target of being in the forefront of product identification technology, well ahead of competitors. It then set out to build up a strong customer base which was used to provide guidance on future developments.

Management

To meet demanding specifications and schedules, effective project management was employed.

Money

The company is 'strong' on financial control, the meeting of production deadlines and the importance of knowing the implications of government policies.

Manpower

The company's success has been attributed to the enthusiasm, drive and commitment of those who work for the company. It is also true that the staff have been motivated by means of appropriate incentives. In practice, the management has to work hard to maintain a balance in the four elements, as the demands in any single one at a given time can be quite extensive. For example, once markets were identified it was necessary to find the manpower to develop them, to control finance and to meet deadlines, through careful organization.

Present status

Donprint's core business has now extended to include security, tracking, traceability and network communication systems. To stay in the forefront the company will be incorporating new advanced materials into its systems, and also sophisticated equipment based on lasers and photochromic dyes.

The current business strategy of the company is aimed at obtaining approximately 25 per cent of the European market sector by 1995. This will require setting up four strategically located production facilities in Europe to support marketing and sales effort. The long-term aim is to achieve a dominant position in North America by the year

2000. To achieve these objectives, Donprint must continue to extend its markets, and increase its money, manpower and management effectiveness in a balanced manner.

7.4 A RESEARCH AND DEVELOPMENT COMPANY

Background

The content of this example is based on 'Organization and Management of R&D in a Privatized British Telecom', the paper presented as the Royal Society Clifford Patterson Lecture in July 1989 by A.W. Rudge, managing director of British Telecom's Research and Technology Division (RTD), together with the subsequent discussion.

RTD (also known as British Telecom's Central Laboratories) is of particular interest as, both before and since privatization, each division of British Telecom (BT) has operated as a subsidiary entity within the parent company. Until 1985 BT was a nationalized company with a monopoly of telecommunication business in the UK and its Central Laboratories undertook work similar to that of a national R&D organization. They provided technical support to the Government, for example, and also cooperated with UK industry to develop products and equipment which were later sold to British Telecom for carrying out some of its services.

Basic questions

- What would be the greatest difference faced by BT's Research Technology Division in changing itself from a nationalized to a private company?
- Who would be the clients of a privatized RTD?
- What objective and strategy should RTD adopt?
- Where would RTD find funds for carrying out research and development projects?
- What changes could the engineers in RTD expect?

Once more it would be useful to prepare your own answers to these questions before reading the text.

Case data

To give a picture of the level of RTD's business, the financial year 1988–89 has been selected as typical. During that year the division's budget was around £200 million, which is 2 per cent of BT's total turnover of approximately £10 billion. Of this £200 million, Group Headquarters provided an allowance of £36 million from corporate funds—usually earmarked for longer-term research and project work—and the balance was 'earned', mainly through development investigations for BT's trading divisions. Two-thirds of the budget was spent within the Central Laboratories, the remainder being used either for work done for BT's trading divisions or for sponsoring external activities such as specific university research projects.

RTD has a total of 4000 scientists and engineers, technicians and other support staff, who were involved in 500 projects in 1988–89. Nearly three-quarters of them work at Martlesham Heath, in Suffolk, while the rest are located at installations in Belfast, Glasgow, Ipswich and London which specialize in computer softward development.

When BT was privatized in 1985, RTD's close relationship with both government and industry was altered, most significantly in the following ways:

- Funding for any research and development initiatives has now to be generated by BT itself, either alone or in partnership with other organizations.
- RTD has had to learn how to compete for business with other companies in the same industrial sector in both the national and international arenas.
- BT has had to modernize its methods and begin developing new products for the future.

We shall now consider how RTD has been coping with the challenges of the new situation.

Performance analysis

With the aid of information given in Dr Rudge's paper and the annual reports of BT, the performance of RTD since the privatization of BT can be analysed under the headings of our four fundamental elements.

Markets

In the past, BT's Central Laboratories did not have to undertake marketing because their role was similar to that of a national research institution and this division was to a large extent insulated from the commercial world. Since privatization, RTD has had to identify its customers and determine their needs. Thus it has had to become directly involved in marketing.

After privatization, RTD's customers could be classified into two groups as follows:

- The operating and trading divisions of BT which, in the year 1988–89, comprised the following: Inland Networks, Customer Services, Business Services, Customer Equipment, International Networks and Services, Mobile and Other Services

- Corporate headquarters

As can be seen, this list of clients includes no outside organizations and RTD's effort takes no direct account of the requirements of British industry, the UK Government or the national interests of the UK, often called 'UK Limited', except where BT's interests and these requirements coincided.

A large number of services to customers are possible, but the effect of privatization on the range was to reduce it to certain key activities, which included:

- Technical support in a number of areas in which BT is active such as computer networks
- Design, development and testing of advanced products and services for telecommunications
- Identification and development of enabling technologies in advance of customers' needs
- Support information for corporate decision making purposes
- Contributions to the identification of fresh technical opportunities, and threats from competitors
- Provision of consultancy services to overcome specific problems

Management

In order to respond effectively to customers' needs, it is important for a management to adopt the most appropriate policy and strategy. Since BT is involved in 'leading edge' technological services and 'high-tech' industrial activities the question of appropriate policy must be clearly reflected in the objectives of RTD. Its basic objective has therefore been stated as follows:

> To provide the defined technological services efficiently and cost-effectively in a commercial and competitive environment.

Dr Rudge described the strategy adopted to achieve this objective as that of an admiral directing his fleet to a promised land. It is surrounded by 'regulatory' shallow water and 'political' rocks, beneath 'public' storm clouds, through a 'fog' of market forecasts which are never totally clear-cut, and against 'winds' of unfavourable economic developments and 'tides' of technological change.

In such circumstances it was felt that RTD's organization must allow customers ready access to its facilities and capabilities, and enable the staff in the Central Laboratories to respond positively and quickly to customer needs. The decision was therefore taken to replace the hierarchical management structure of RTD with a 'matrix management organization' whereby individual departments provide a resource base of people and facilities and all assignments would be handled as departmental projects. This means that a high degree of project management capability is needed by all those who will take charge of projects. Particular emphasis is placed on the ability to produce the required results on time and within the agreed budget.

Other important aspects of the management function include efficient management of information so that all projects in progress at a given time may be interlinked, effective use of negotiating skill to achieve the optimum work arrangement for projects, and careful direction of supervision of research and development effort to ensure that they are indeed implementing the stated policy.

Money

As already indicated, RTD's revenue is generated within the divisions of BT and in order to offer cost-effective services to customers this division has to have strong financial management control and also sound criteria for measuring performance. As things now stand it is not possible to charge the 'under-utilized' direct cost of researchers' time to the customers as part of an increased amount for overheads. In practice, these may include indirect costs such as staff involved in administering the project, the depreciation of equipment and facilities and the supply of miscellaneous items. It is therefore essential to make the most efficient use possible of technical staff time through more flexible organization.

To meet the terms of RTD's basic objective, each department has been required to adopt sound methods of financial control including the preparation of a business plan each year and providing a monthly measurement of financial performance.

With regard to RTD's budget level, in 1988–89 this represented 2 per cent of BT's annual turnover. Whether this is the 'correct' percentage can only be determined by comparing BT's practice with that of other business enterprises while taking into consideration its own special circumstances. It is worth noting that in the commodity products industries, research and development budgets usually represent between 0 per cent and 5 per cent of the turnover while levels of 6–20 per cent can be found in companies making high-technology products. It may take quite a time to establish the correct level for RTD during the 1990s. The final figure will be affected by factors that include the state of the world economy, the policies of governments, the effectiveness of the Division's financial management and the performance of BT as a whole.

Manpower

The aspiration of Central Laboratories is to be the 'engine of change' within BT. But to achieve this it was realized that the attitude of staff needs to be altered at every level, and in particular at the senior management level. These people are expected to be 'not only very good technologists but also fully rounded business managers'. In order to fulfil the objectives of the organization it is necessary to provide the means of staff development for technical managers so that their performance will be improved.

RTD employs around four thousand people who are expected to be technologically competent and able to respond positively to a variery of demands. This requirement does highlight the importance of good communication skills among the staff. These skills are essential for example, for effective identification of customers' needs, and for ensuring that work is done efficiently and punctually. They are exercised in a number of ways, including oral presentations, interaction with team members and customers, written reports, and elucidation of complex technological issues for colleagues in management.

With the strong emphasis that is laid on the project management approach to assignments in RTD, all its scientists and engineers must have a good appreciation of the key features of this approach and be able to act effectively as either a team member or a leader.

It is possible for human resource, like equipment and facilities, to become 'obsolete' unless the capabilities of staff are continually 'maintained and upgraded'. The work of bringing this about should not be confined to formal training course of varying lengths, but should include informal aspects such as learning on the job and self-development. Important among the attributes required in staff are the ability to understand the value of both their own time and that of other people, and skill in making the most effective use of it. These contribute directly towards the cost-effectiveness of the services provided by RTD.

Comments

Although it may be early to draw firm conclusions on the success or otherwise of RTD, it is already learning to cope in a very competitive commercial world. In this connection a number of points deserve consideration and these will again be examined under the four main headings.

Markets

The staff of BT's Central Laboratories are now more market-orientated. It should be noted that the majority of their customers are not interested in buying research but want, and will pay for, solutions to specific problems. If it emerges that research is needed at a particular price in order to achieve that solution, they will pay that price. This is quite different from the position when the company was nationalized. It is a fact of life that the researchers, however enthusiastic, would find it extremely difficult to persuade a customer to buy work that is not really required.

Management

Having to operate in such competitive and rapidly changing conditions, it is essential for management to instil into the thinking of all its staff an awareness that uncertainty is actually the 'norm' for the business. In addition, in the case of RTD, the management has to be prepared to respond very quickly in its own operating environment. The managers of individual projects are expected not only to be strong technically but also to be highly skilled in balancing, as well as is practically possible, the demands on:

- Time
- Money
- Technology
- People

Money

Two issues deserve attention here. Firstly, expenditure on research and technology, which represents 2 per cent of total turnover, appears to be very low for a major telecommunications business. If this level is maintained, and additional resources are needed to meet specific research objectives, e.g. the development of future enabling technologies, it may be necessary to engage in joint ventures, either nationally or internationally, with other companies or with government organizations.

Secondly, the cost-effectiveness of the service could be increased by better utilization of human resources, but great care needs to be exercised to ensure that this is properly interfaced with changes in customer demand and in technology.

Manpower

The background information did not really address the question of human resources in any depth except to highlight the importance of having a pool of high-quality managers with a good level of technological skill and of project management ability. This is a matter of extreme importance because there is a danger that BT's Central Laboratories will not have sufficient technical personnel to cover the work involved in the number of research and development tasks being undertaken.

Present status

The revision of policy and reorganization have now taken place and time is needed to determine whether they will be successful. RTD has made an excellent start by making the needs of the customer the top item in its list of priorities. It is also learning to identify the most appropriate level of research and development expenditure and the 1988–89 figure of 2 per cent of the parent company's turnover can be expected to change in due time. Criteria are being established for measuring and monitoring the transfer of techniques, methods, products and service ideas from the laboratories to the customers. This latter process has now been established and applies to both business done between the Central Laboratories and the operating divisions and also inside the Central Laboratories.

The effect of implementing this policy has been to emphasize the need for the laboratory work programmes to be continually revised in response to customer demand. For example, the current huge growth in computing is leading to software system engineering playing an increasingly dominant role in the activities of BT's operating divisions.

It is absolutely correct that the funds sought by the Central Laboratories should be used to satisfy customer needs rather than for work that the researchers want to do. However, a proper balance must be kept between customer needs and preparation for the future. The work programme should be reviewed regularly to ensure that some of the longer-term requirements, such as developing enabling technologies in anticipation of future needs, are not forgotten in the rush to meet pressing short-term demands.

CHAPTER 8

APPLICATION

8.1 TYPES OF APPLICATION

The previous five chapters have considered the key aspects of Markets, Management, Money and Manpower together with case examples. We shall now illustrate how these fundamentals of business can be applied positively in the practical situations likely to be encountered by the engineer.

As indicated in Section 2.5, there are two broad types of application, which are entitled:

- Appreciation
- Further studies

As these titles imply, the first kind of application involves using a knowledge of the fundamentals of business in order to understand the work of those in management and to improve interaction with other technical personnel. As a result, the engineer's professional work will be enhanced and interaction with other professions can also become more effective. The second type of application involves using this fundamental knowledge as a basis for further study of certain selected topics in order to achieve a high degree of proficiency in them.

Before examining these two forms of application in more detail, it would be useful to outline the assumptions employed here. Firstly, it is assumed that the engineer's work time is basically devoted to his or her professional subject and thus only a limited number of subjects can be studied in greater depth. Study of this kind may be done either informally by reading books and other publications or formally by attending suitable courses. Typical examples of such courses would be a series of classes in marketing offered by an educational institution or a three-day programme on project management organized by a company specializing in short courses.

Secondly, the engineer requires different aspects of business fundamentals during each phase of his or her career. For example, it is more important for a student to acquire skill in teamwork than to make a detailed study of methods of obtaining finance for commercial projects.

Thirdly, the learning process should be a continuous and 'dynamic' one. The information gained and skills acquired through study should be put into practice at the earliest possible opportunity.

Bearing these points in mind, the two types of application will now be considered.

8.2 APPRECIATION

In this type of application, the emphasis is on gaining a basic appreciation of business fundamentals and on making the best use of them. Aspects deserving special attention are described below.

Becoming familiar with the language

Like any other subject, each business activity has its own terminology and jargon. By becoming familiar with this language engineers can better appreciate the usefulness and the limitations of these activities. They should be greatly helped in this respect by a study of this text and the glossary provided, although it would not be possible to understand all the terms or jargon unless they were being regularly used in practice. However, a grasp of the language of business will certainly assist communication between the engineer and those trained in business subjects.

Understanding the key issues

The practical application of business techniques can be highly complicated because it involves many factors, details and human interactions. However, the fundamentals are expected to remain fairly constant, so much so that they are often forgotten even by those who have specialized in this area. Accordingly, if the engineer has grasped the salient feature of each component of business, he or she should be able to apply this understanding so as to place the correct emphasis on each one. Typical examples of this are the way in which an understanding of the goal of selling can lead to a fresh approach to the engineer's daily work, and the way in which an appreciation of what project management is trying to achieve can improve an engineer's overall efficiency.

Asking the right questions

A good basis of knowledge will help to ensure that the right questions are asked of everyone involved in the different aspects of an activity. As an example, the engineer responsible for overseeing a project would expect to find the following information in a project proposal:

- Objective of the project
- Strategy to be adopted for achieving the objective
- General review and market research results
- Information for planning the project
- Sources of funds and a forecast of their expenditure
- Composition of the team and the choice of leader

He or she should be raising appropriate questions if any of this information is missing.

Making judgements

Many assignments and project ideas have to be assessed for feasibility and in doing this technically trained personnel tend to focus more strongly on the technological factors. An appreciation of the fundamentals of business, however, will enable them to take both technological and commercial criteria fully into consideration. This in turn is likely to lead to better decisions and hence increase the chances of a successful operation.

8.3 FURTHER STUDIES

There are two ways of building on a knowledge of the fundamentals of business and using it as the basis for further study. The first has been called the 'systematic approach', and this assumes that additional knowledge and skills can be gained and employed systematically throughout the career of the engineer. The second, termed the 'random approach', assumes that the need for particular knowledge and skill will arise in relation to a given situation and at that point the subject is studied in order to cope efficiently with the job. These two approaches will now be considered in more detail.

The systematic approach

In this approach the career of an engineer is divided broadly into the following four phases:

- *Student phase*:   The period when a full-time or part-time course in engineering is being followed
- *Graduate engineer phase*:   This period covers the first five years after graduation from an educational institution
- *Section leader phase*:   This is the stage when the engineer is given greater responsibility
- *Manager phase*:   This covers a period when a more senior management role is taken in an organization

It is suggested that sufficient time is allocated for taking a maximum of six topics to greater depth during each phase. For this purpose each of the six should be assigned either primary (P) or secondary (S) priority, as suggested in Table 8.1.

The random approach

This approach is suitable for the many practising engineers who did not study business topics or become involved in the 'business' activities of their organization until well into their careers. There are various reasons for this state of affairs, typical examples being a lack of interest on the part of the individual concerned, the technological emphasis of the education programme undertaken or the sheer difficulty of using the published books on the subject. Having gained a basic appreciation of the fundamentals of business, such engineers can enhance their understanding of this whole area by selecting a limited number of topics to be studied to greater depth. To illustrate this approach, four broad groups of engineers have been selected as follows:

- *Entrepreneurs*:   Those who wish to run their own business
- *Consultants*:   Those who work in consultancies or similar organizations
- *Researchers*:   Those whose work is concerned with research and development
- *Academics*:   Staff in educational institutions with teaching, research and administrative responsibilities

In each case it is assumed that there is sufficient time or interest for a closer examination of up to eight topics and, as before, each set is divided between primary and secondary priorities, as summarized in Table 8.2.

Table 8.1　A summary of the systematic approach

Subject	Career phase			
	Student	**Graduate Engineer**	**Section Leader**	**Manager**
3.1 Basics of marketing	S			
3.2 Market research		S		
3.3 Opportunities and timing				P
3.4 Quality and competition		S		
3.5 Effective selling			S	
3.6 Innovation and invention				S
4.1 Objectives and strategies	P			
4.2 Dealing with contracts		P		
4.3 Approach to negotiation			P	
4.4 Project management	S			
4.5 Information management			S	
4.6 Research and development				S
5.1 Understanding economics	S			
5.2 Sources of funds			S	
5.3 Accounting methods		P		
5.4 Performance assessment			P	
5.5 Assessment of project viability				P
5.6 Government policies		S		
6.1 Human resources				P
6.2 Communication skills	P			
6.3 Working as a team		P		
6.4 The leadership role			P	
6.5 The role of education				
6.6 The organization of time	P			S

Table 8.2 A summary of the random approach

Subject	Type of career			
	Entrepreneur	Consultant	Researcher	Academic
3.1 Basics of marketing			P	P
3.2 Market research	P	P		
3.3 Opportunities and timing	S			S
3.4 Quality and competition		S		
3.5 Effective selling				
3.6 Innovation and invention			S	
4.1 Objectives and strategies	P	P	P	P
4.2 Dealing with contracts				
4.3 Approach to negotiation	S			
4.4 Project management		S		S
4.5 Information management			S	
4.6 Research and development				
5.1 Understanding economics	P			
5.2 Sources of funds			P	S
5.3 Accounting methods				
5.4 Performance assessment		S		P
5.5 Project viability	S			
5.6 Government policies		P	S	
6.1 Human resources				S
6.2 Communication skills	P	P	P	
6.3 Working as a team		S		P
6.4 The leadership role				
6.5 The role of education			S	
6.6 The organization of time	S			

Glossary of Common Business Terms

Accounting period   The period of operation covered by one set of company accounts. This is usually one year.

Asset   An item of specific value to a company, e.g. land, buildings, machinery, cash etc.

Balance sheet   A statement of the assets and liabilities of a company at a given point in time, usually the end of a trading period such as one year.

Base rate   A rate of interest used as reference by financial organizations when setting their own interest rates for investors and borrowers. It is set by the Government via its principal banking organization. In the UK, for example, it is the Bank of England that determines this rate.

Capital   The amount of finance available or committed for some aspects of a business activity, such as starting up or undertaking a new project.

Cashflow   The sum of money represented by the difference between income and expenditure.

Charity   An organization which has been set up for the sole purpose of providing support for non-profit-making activities.

Competition	The contest for customer preference between rival organizations offering similar products, services or functions.
Contract	A formal agreement between two or more parties which is usually enforceable by the laws of a particular country.
Current assets	Assets which a company holds but may dispose of within a year.
Deadline	The specified time point by which all work on a particular task has to be completed. For example, the deadline for receiving proposals for work on a new project is often noon on a given date. None that arrive after that time will be considered.
Development	The process that takes the most commercially promising research results forward to the stage of practical implementation so that, for example, a product can be put on the market.
Discount	A deduction from some standard or reference value. For example, if a product listed at £100 is sold for £90, this means a 10 per cent discount has been applied, or a discount rate of 10 per cent.
Dividend	A payment made to the shareholders of a company out of its profits.
Earnings	The income derived from the operations of a company.
ERM	The 'Exchange Rate Mechanism', which is a method of controlling fluctuations between the exchange rates in the European Community by allowing only a small percentage variation around the reference currency.
Fixed assets	Assets which a company plans to hold for at least one year.
Forecast	A prediction of likely future trends and developments on the basis of certain assumptions.
Foundation	An organization responsible for investing a certain sum of money and distributing the income for the support of research or development activities, or charities.
Funds	The money required by an organization to support its activities.
Grant	Usually a specified amount of money which does not need to be repaid, given by an awarding organization to an individual, company or institution for a particular purpose.
Information	Any collection of facts, figures, knowledge, relationships, experience, decisions or assumptions.

Initiative   A venture for which eventual success is hoped.

Innovation   Improvements to an operation, system, piece of equipment, or production procedure, achieved through a fresh use of existing or known technologies or methods.

Interest   The amount of money, expressed in percentage terms, to be paid by borrowers to investors for the use of a sum of money over a specified period. For example, a company borrowing £1000 from a bank for a three-month period at a rate of 20 per cent would have to pay £50 interest in addition to returning the borrowed £1000.

Invention   An original idea for a product or process, conceived and physically realized by a completely new treatment which is not an obvious extension of existing or known technologies or methods.

Investment   A sum of money put into a potentially profitable venture in the hope of increasing its value in the future.

IRR   Internal Rate of Return. This is a method for evaluating the discount rate so as to determine whether an investment is worthwhile.

Loan   A sum of money given temporarily to one individual or organization by another, to be repaid, usually with interest, after a certain period of time. For example, a bank could lend £50 000 to a company for a period of three years.

Loss   The shortfall between the expenditure on an operation and the amount of income generated by it during a given period.

Market   The location, environment or source of potential customers for goods or services on offer.

Marketability   The degree to which a product or service is attractive to potential customers. A highly marketable product is one which is in great demand while a product with low marketability is one for which customer demand is low.

Market forces   The effects on the demand for a product, service or action generated by the preference of the customers or those involved in the activities concerned. This preference can alter rapidly within a very short space of time.

Marketing	Activities associated with identifying the needs of existing and/or potential customers, anticipating when these needs will arise and ensuring that they can be satisfactorily fulfilled.
Market research	The gathering and analysing of information in order to forecast future commercial opportunities on the basis of certain assumptions.
Merger	Two companies combining their assets, expertise and staff in order to operate as a single organization.
Milestone	The point in time when the current achievements of a project or business are to be reviewed against its objectives and planned work programme.
Monopoly	The situation in which one company holds an exclusive or dominant position in a particular commercial operation or activity.
NPV	Net Present Value. The calculated value in today's terms of probable income at a future date from a given investment. This calculation is done in order to compare the potential viability/profitability of projects competing for the same resources.
Objective	The target which defines the overall direction to be taken by an organization.
Opportunity	A situation which, if exploited properly and at the most suitable moment, could yield success.
Payback method	A method for assessing the viability of a project based on calculation of the time taken for the cashflow to be equal to the capital investment.
Profit	Usually the excess of a company's income over expenditure for a given period. Profit considered before the deduction of taxation and other liabilities is described as gross profit. The amount remaining after these items are deducted is referred to as net profit.
Profitability	The likelihood of a business being able to generate a profit through its operations.
Quality	The term used to define a product or service which very closely meets a stringent set of agreed specifications, or shows the greatest fitness for its purpose.

Ratio
The measurement of one value against another, often in a non-dimensional form. In business terms, the ratio of, e.g. profit to turnover, provides a useful means of comparing the performance of one company with that of another.

Repayment
The return to the lender of a borrowed sum of money. This can be done in a single sum of money, or by means of a series of payments at agreed intervals.

Research
Study or investigation directed at increasing knowledge and understanding of an idea, concept, problem or phenomenon.

Reserves
The portion of profit after taxation which is not paid out as dividends but is set aside for re-investment in the company.

Selling
The process whereby one party persuades an individual, an organization or the public at large to accept an idea, product or service in exchange for some reward, whether money, a positive response, agreement or general support.

Share
A financial 'stake' in a business venture, entitling the holder to a say in policy decisions and to a proportion of the distributed profits.

Start-up
The period when a company is commencing its business activities.

Strategy
A series of logical steps to ensure that an objective, for example of an organization, can be achieved.

Subsidiary
A company whose operations are effectively controlled by another company, for example, through holding more than 50 per cent of its share capital.

Subsidy
Financial or other aids given to a business, either by the government or by another company in the same group.

Takeover
One company gaining financial control of another.

Trade
The exchange of goods for money or other goods, whether between individuals, organizations or nations.

Turnover
The total income of a company over a given period of time such as one year. The turnover figure indicates the level of a company's business activities and is a useful factor for comparison with other companies involved in similar activities.

INDEX